A Theory of Philosophical Fallacies

Argumentation Library

VOLUME 26

Series Editor

Frans H. van Eemeren, *University of Amsterdam, The Netherlands*

Editorial Board

Bart Garssen, *University of Amsterdam, The Netherlands*
Scott Jacobs, *University of Illinois at Urbana-Campaign, USA*
Erik C.W. Krabbe, *University of Groningen, The Netherlands*
John Woods, *University of British Columbia, Canada*

More information about this series at http://www.springer.com/series/5642

Leonard Nelson

A Theory of Philosophical Fallacies

 Springer

Leonard Nelson
Göttingen
Germany

Translated by Fernando Leal and David Carus

Deceased—Leonard Nelson (1882–1927)

ISSN 1566-7650 ISSN 2215-1907 (electronic)
Argumentation Library
ISBN 978-3-319-37323-2 ISBN 978-3-319-20783-4 (eBook)
DOI 10.1007/978-3-319-20783-4

Springer Cham Heidelberg New York Dordrecht London
© Springer International Publishing Switzerland 2016
Softcover reprint of the hardcover 1st edition 2016
Main text translated from the German language edition: *Typische Denkfehler in der Philosophie* by Leonard Nelson, © Felix Meiner Verlag 2011. All rights reserved

Appendix translated from the German language edition: "Die kritische Ethik bei Kant, Schiller und Fries: eine Revision ihrer Prinzipien", *Gesammelte Schriften in neun Bänden*, vol. VIII, pp. 27–192 by Leonard Nelson, © Felix Meiner Verlag 1971. All rights reserved

This work is subject to copyright. All rights are reserved by the Publisher, whether the whole or part of the material is concerned, specifically the rights of translation, reprinting, reuse of illustrations, recitation, broadcasting, reproduction on microfilms or in any other physical way, and transmission or information storage and retrieval, electronic adaptation, computer software, or by similar or dissimilar methodology now known or hereafter developed.
The use of general descriptive names, registered names, trademarks, service marks, etc. in this publication does not imply, even in the absence of a specific statement, that such names are exempt from the relevant protective laws and regulations and therefore free for general use.
The publisher, the authors and the editors are safe to assume that the advice and information in this book are believed to be true and accurate at the date of publication. Neither the publisher nor the authors or the editors give a warranty, express or implied, with respect to the material contained herein or for any errors or omissions that may have been made.

Printed on acid-free paper

Springer International Publishing AG Switzerland is part of Springer Science+Business Media
(www.springer.com)

Contents

Introduction by Fernando Leal 1
What Is the Theory? 2
Who Was Leonard Nelson? 6
Some Objections to the Theory 9
Note on the Translation 13
References 18

Lecture I 21
References 28

Lecture II 29
Reference 34

Lecture III 35
References 41

Lecture IV 43
References 50

Lecture V 51
References 56

Lecture VI 57
References 63

Lecture VII 65
References 71

Lecture VIII 73
References 81

Lecture IX	83
References	89
Lecture X	91
References	98
Lecture XI	99
References	107
Lecture XII	109
References	116
Lecture XIII	117
References	125
Lecture XIV	127
References	134
Lecture XV	135
References	142
Lecture XVI	143
References	149
Lecture XVII	151
References	158
Lecture XVIII	159
References	165
Lecture XIX	167
References	174
Lecture XX	175
References	181
Lecture XXI	183
References	190
Lecture XXII	191
References	200
Appendix: Seven Kantian Fallacies	203

Introduction by Fernando Leal

Abstract In 1921 Nelson presented what seems to be the first theory of philosophical argumentation, or at least of its negative or destructive part—a theory of philosophical fallacies. That theory says that the peculiar nature of philosophical thinking repeatedly leads people to take ordinary concepts with great philosophical import, such as causality or duty, and replace them with new, made-up ones, whose content is very different from that of the originally available ones. When the new concepts usurp the place of the original ones within a philosophical argument, they invariably produce false results—on all sides of philosophical disputes. Many apparently irresolvable disputes in philosophy are the product of such fallacious concept-swapping.

The relationship between philosophy and argumentation is, to say the least, peculiar. Philosophers argue all the time, that much we all know, even though quite often it is difficult, and sometimes exceedingly difficult, to make out what exactly their arguments are. But even conceding (for the sake of argument) that philosophy is all about arguing, it is again a hard fact that philosophers, in sharp contrast to mathematicians or scientists, are rarely if ever convinced by each other's arguments. Moreover, the favourite occupation of philosophers is to pick holes in the arguments of their fellow philosophers, thereby indicating to the outside world that there is something rotten with *all* philosophical arguments.

None of this is controversial. A curious observer would therefore be readily excused if he or she would draw the conclusion that, surely, philosophers *must* have by now developed a theory of philosophical argumentation. This is hardly the case. A careful inspection of the relevant literature reveals it to consist basically of two kinds of publications. On the one hand, there are practical 'textbooks' purporting to give students *some* analytic tools to analyze and evaluate philosophical arguments or to construct and present their own. On the other hand, there are metaphilosophical texts presenting a certain view of the nature of philosophy with more or less loose observations about the *role* that arguments play or do not play in philosophy.[1] None of these books, however, present a systematic *theory* of philosophical argumentation *as such*, a theory of how philosophers actually argue or fail to argue.[2]

This is the main reason why the book the reader is now holding in his or her hands merits attention. This book *does* present a theory of philosophical argumentation. It is perhaps too soon to say whether the theory is correct, entirely or in part, but it is quite a clear theory, so that at the very least it might start a much needed discussion about what a theory of philosophical argumentation should be.

In the following I shall first briefly state what that theory is, then I shall present its author, for he is not well known, and finally I shall endeavour to meet some objections that can be raised against the theory.

What Is the Theory?

Nelson's theory of philosophical argumentation, as set forth in this book, is only a partial theory, for it only concerns those philosophical arguments that go wrong, that contain a fallacy, i.e. an error in reasoning. It does not say anything about philosophical arguments that are correct and fallacy-free. Moreover, the theory does not purport to cover all fallacies, but only the 'typical' ones, viz those that a philosopher *will* commit sooner or later *because* of the peculiar business he is in. Fallacies of such a typical sort are as it were an occupational hazard.

Even the plural—fallacies—has to be taken with a pinch of salt, for at bottom there is, according to the theory, only *one* typical fallacy in philosophical argumentation, although it comes in many shapes and forms. This may not look very promising. A theory of philosophical argumentation that is in fact about *one* fallacy? For it to work we must assume that in philosophy we will, always or at least for the

---

[1]To the first group belong e.g. Cornman et al. (1992), Rosenberg (1996), Morton (2004), Cohen (2004), Martinich (2005), Baggini and Fosl (2010); to the second e.g. Stove (1991), Glymour (1992), Bouveresse (1996), Williamson (2007), Gutting (2009), Schnädelbach (2012). I am naturally excluding from consideration those textbooks written by philosophers about argumentation in general (usually in view of courses in logic or critical thinking) as well as papers or books in which particular arguments of particular philosophers are subjected to scrutiny either for historical or systematic purposes. They may contain valuable remarks about philosophical argumentation but belong to a different ballgame altogether. A special case is Nicholas Rescher, who in a series of papers and books (e.g. 2001, 2006, 2008) has made some interesting, albeit in my opinion somewhat rambling, attempts at a theory of philosophical argumentation.

[2]There seem to be two theories of philosophical argumentation apart from Nelson's. Johnstone (1959, 1978), developed in a long and distinguished career a theory that is both profound and undeservedly underrated by philosophers, although it has fortunately met with some attention in the field of argumentation studies (see e.g. the special issue in his honour in *Informal Logic*, 2001, as well as van Eemeren and Houtlosser 2007). I can see clear points of contact as well as undeniable differences between Johnstone's theory and Nelson's, and I certainly hope to tackle the task of comparing them in the future. As for the other theory, see Footnote 7 of this chapter.

most part, reason correctly *if only* we avoid that one fallacy to whose description and illustration this whole book is devoted. That is exactly what Nelson assumes.

But what exactly is this fallacy? The analysis that uncovers it is a two-step procedure.

Step 1 Nelson starts with the fact that a philosophical argument is always part of an ongoing, if not always fully explicit, discussion between philosophers each one of whom affirms something that appears to imply denying what the other holds. The logical building constituted by the opposing arguments Nelson calls the 'dialectics' of the issue at hand—harking back to Plato, who invented the term to name the peculiar art of dialogue developed by his master, Socrates. Although such logical buildings—the sets of propositions and arguments pitted against each other all along the history of philosophy—can sometimes be abstruse and convoluted, Nelson's analysis tries to reduce the basic opposition to its raw core, which he sometimes represents in the shape of elegant diagrams. The reader will find in this book five of those. Popper for one expressed his admiration for this analytical feat of reduction and tried to imitate it (see Popper 1979).

Those diagrams show that the way philosophers arrive at their positive philosophical doctrines is by way of denying the positive doctrines of their opponents. They manage to believe they have proved something, say P, because they have shown the opposite doctrine, say Q, to be false. Their opponents play exactly the same game, so that the net result is the appearance of a relation of contradiction between P and Q where there is in fact no such contradiction. The game is played again and again by philosophers on the strength of a common assumption, a usually hidden presupposition shared by the two adversaries, viz that P and Q exhaust all possibilities available to philosophical thought. Yet if P and Q exclude each other, their negations do not. In other words, if P is true, then Q has to be false, and if Q is true, then P has to be false; but if P is false, Q does not have to be true, and if Q is false, P does not have to be true. In the traditional logical jargon, the two philosophical doctrines are not *contradictory* but rather *contrary* to each other—they cannot both be true yet they may very well both be false, viz if the common assumption that they are the only alternatives is itself false. The starting point of the discussion is then the choice between two options (either P or Q), a logical disjunction which is understood by both discussion partners as being both exclusive *and exhaustive*. Starting from such a disjunction the discussion evolves in a predictable way, which Nelson's diagrams try to represent in a terse, parsimonious manner. If we use arrows to indicate an inferential relation, then the following would be the general form of a Nelson diagram:

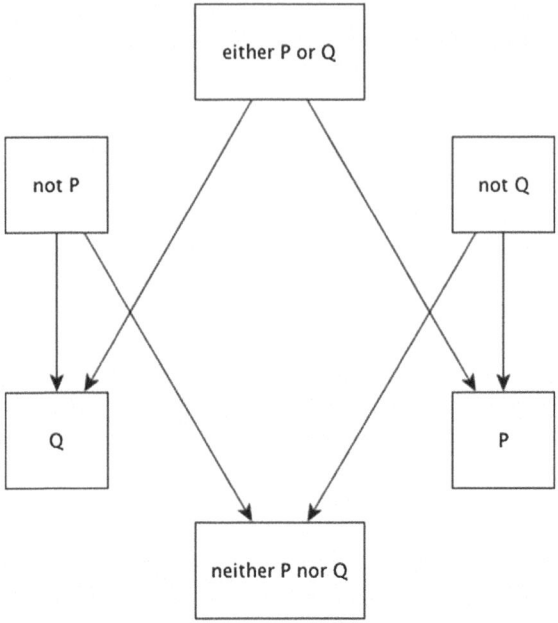

We see here that there is contradiction between *P* and not-*P* as well as between *Q* and not-*Q*, but there is no contradiction between *P* and *Q*; not even if we concede that *P* and *Q* exclude one another. At least there is no contradiction unless and until we assume that *tertium non datur*, that there is no third possibility, say *R*, which excludes both *P* and *Q*. Nelson's argument is not only that *R* is available but that *R* is indeed the true alternative.

So far we would have one form of what is usually referred to as a false dilemma or a false dichotomy (Nelson himself calls it an 'incomplete disjunction'). It would not be much of an exaggeration to say that for Nelson the history of philosophy has a dilemmatic structure, and that all breakthroughs in that history consist in showing some of these dilemmas to be false. But the question now arises as to why philosophers become prisoners of a false dilemma in the first place. This leads to the second, and the most important, step in Nelson's analytical procedure.

Step 2 This consists in breaking down the sentences and arguments used in philosophical disputes so as to identify the underlying concepts. The leading idea is that quite often at least one of those is a fake and a phony, an extra-ordinary concept masquerading for an ordinary one by an artful or unwitting sleight of hand, something that a philosopher has *made up* instead of respecting the meanings our common vocabulary puts at our disposal. Such a replacement of one concept by a different one—sometimes introduced by a seemingly harmless definition—creates an instance of equivocation, or what in traditional syllogistic was called *quaternio terminorum*, a well-known fallacy by which the same word or phrase is used, within the same argument, in more than one meaning. When this happens, a false dilemma is automatically and inevitably produced which the disputants are often unaware of. If an analogy is allowed, fallacies in philosophy are, according to Nelson's theory, very much like accidents—they are harmful, nobody wants them to happen, and yet they happen because people do not attend to the business at hand in the proper manner. Or again, to commit the 'philosophical fallacy' is very easy, in fact so easy that it is in a sense surprising that philosophers do not commit it even more often.

Let us look at this second step more leisurely. To start with, it is very important to note that card-carrying professional philosophers are not the only ones that produce philosophical arguments. It is certainly their business to do so, and they do so much of it that the pool of arguments philosophers have produced is the most bountiful source of examples for a theory that concerns itself with philosophical fallacies. Nonetheless, philosophical laypersons also occasionally develop philosophical arguments. Physicists such as Einstein or mathematicians such as Poincaré did; and so does each one of us, especially in extreme situations. Who has not indulged in a bit of philosophising after losing a loved one or when everything seems to go wrong or as one faces a difficult moral dilemma? Philosophical argumentation, and with them philosophical fallacies, are to be found outside the writings and conversation of professional philosophers as much as in them. For it is the business itself of doing philosophy, no matter who gets involved in it, that is accident prone in the sense of that one fallacy Nelson theorises about.

Consider the case of an ordinary person in a stressful situation. If we set aside the ancient consolations of religion and the allure of modern psychotherapy, what people have at their disposal to face death, disease, betrayal, and other hard facts of the human predicament is a culturally given set of ready-made words (e.g. 'fate', 'life', 'honour', 'deserve', 'true'), phrases (e.g. 'bad luck', 'reasonable doubt', 'morally certain', 'social justice', 'as good as it gets'), and sentences or half-sentences (e.g. 'what goes around comes around', 'better safe than sorry', 'what doesn't kill you makes you stronger', 'you had it coming', 'life is unfair'). These pieces of vocabulary are for the most part of unknown origin—the product of accumulated wisdom or folly—albeit a few of them can admittedly be given a date and sometimes even an author's name. We use them to think through our troubles, usually in more or less anguished conversation with friends or relatives, sometimes in solitary brooding. With them we construct our arguments, sometimes correctly, more often not.

Saint Augustine famously declared once about time: 'If no one asks me, I know what it is. If I wish to explain it to him who asks, I do not know.' That is exactly what happens with every single piece of the vocabulary we use when we try to find our way out of or around our philosophical conundrums. If nobody challenges us, we sail with the wind; but if someone expresses a doubt, or if we ourselves have doubts, we are stuck. One way we try to respond to challenges and doubts is by committing *the* fallacy. What we do then is to replace the meaning of whatever piece of vocabulary we are using with a very special and different concept or idea, the consequence of that concept-swapping being that one can now prove practically anything.

Once the vocabulary has served its purpose in allowing us to cope with our present difficulties, we continue with the ordinary business of life—leaving behind with alacrity the extra-ordinary business of philosophy. For make no mistake: in those few moments we are all philosophers and we all do what philosophers do—try to make sense of things by means of a given vocabulary that has emerged from uncountable past human interactions for the special purpose of understanding and orientation in difficult situations. Unlike professional philosophers, most people are indeed not particularly attracted to philosophy. They find it unhelpfully complicated and largely a waste of time—except in troubled moments. Because of this, a theory of philosophical argumentation had better concentrate upon the writings and sayings of those who make philosophy their calling. Nonetheless, the reader should understand that the fallacy Nelson singles out as the most typical is not a sad failing of a particular group of people—*the* philosophers—but an accident waiting to happen to anyone who does occasionally what they do routinely.

Coping with a present tribulation is just an example of how anybody comes to philosophise—or to commit the fallacy. Sometimes it is when we have narrowly escaped a mishap or even when we have been granted a great and unexpected blessing that we start philosophising. And the temptation to indulge in philosophical concept-swapping can overpower us if we want somebody to acquiesce to our wishes. When children or teenagers are trying to make a parent do what they want, or allow them to do something, they often bend certain words in particular ways. Thus 'fairness' will be made to mean equality if equality is what leads to the desired conclusion, but otherwise it will be made to mean the opposite of equality. As for philosophers, we all know that they become famous because they have a powerful imagination that allows them to build complicated and often overwhelming worldviews—sometimes just giving expression to popular prejudice (e.g. nationalism, naïve realism), sometimes actually creating a startingly new way of looking at Life and the Universe (think of Plato or Descartes). And they come to believe in their worldviews so strongly that they use whatever arguments they can come up with. If Nelson is right, equivocations and false dilemmas often are the consequence of such wishful thinking.

So that is the theory. Its building materials are nothing new, in fact they are exceedingly well known. Equivocation was already identified by Aristotle over two thousand years ago; and one of the clearest examples of an error in reasoning is doubtless the false dilemma (for a little-known contribution to show how widespread this fallacy is, see Vaz Ferreira 1916). And yet Nelson manages to construct a bold theory of philosophical argumentation with those humble materials. As long as a philosopher is careful and avoids assigning an arbitrary, made-up concept to the ordinary words in the dialect of his tribe, he will not fall victim of disputes based on false dilemmas, and so he will reason correctly.

Who Was Leonard Nelson?

Born in Berlin in 1882 into a very well-to-do and highly cultured family, Leonard Nelson managed to write, starting in 1904, well over one and a half million words of closely argued philosophical prose, and to teach many courses at the university whilst leading a group of like-minded and tremendously disciplined young people with whose help he edited books and periodicals, founded a school both for children and for the training of political cadres, and actively participated in the social and political life of the Weimar Republic both within the German Social Democratic Party and within autonomous organisations upholding a non-Marxist version of socialism, before he died, at 45 years, of exhaustion and pneumonia in 1927. Very few philosophers can claim to have done so much and so competently in so little time.

Nelson's academic writings run all the gamut from mathematical logic and the philosophy of mathematics through epistemology and the philosophy of natural science down to ethics, political philosophy, the philosophy of law and the philosophy of education; his popular writings cover philosophical matters as well as the day-to-day social, cultural and political life of Germany and Europe. There are many ways in which one could describe Nelson, but the best summary might be to say that he was an untimely philosopher. An orthodox Kantian[3] in an age in which Kant's doctrines were being either adulterated or thoroughly repudiated, Nelson was also an analytic philosopher of the purest breed as well as the clearest and tersest of philosophical writers in a country and a language that were then thoroughly opposed to such ways of thinking and writing. And to crown it all, he actually believed that the purpose of one's ethical and political thinking was not to show off one's cleverness but rather to find out how one ought to live, and having found out, to go on and change one's own life and public conduct accordingly (see Leal 2013).

Yet, in spite of such variety, one motif pervades the whole of Nelson's philosophical œuvre, a motif common to all early modern European philosophy, viz the search for the right method to do philosophy and its relentless application. He firmly believed that—standing on the shoulders of Bacon, Descartes, Newton and Hume, but correcting their mistakes—Kant was the first to arrive at a proper formulation of the right method in philosophy. Although basically right, Kant's formulation, according to Nelson, needed some polishing and further development. In fact, for him as for Kant the phrase 'critique of reason' was emphatically not the name of a book but the name of a 'new science', to use Galileo's term. And so as the new sciences created by Galileo—those we now call 'kinematics' and 'material mechanics'—have gone a long way since Galileo, do not depend on his name and authority to keep developing, and in fact are the fruit of the work of generations of researchers, so should the 'critique of reason' be as conceived by Nelson—a humble scientific discipline, no more and no less. Not to put too fine a point on it, the 'critique of reason' is in this conception nothing but a branch of what we now call 'cognitive science', viz the branch dedicated to explore the conditions or

---

[3]Orthodox but not dogmatic. Nelson has objected to some of Kant's arguments using the very same method he applies to everyone else. To prove this, I have added, at the end of the present book, an appendix in which our author illustrates no less than seven ways in which Kant did fall for the typical philosophical fallacy.

principles underlying our various forms of knowledge and action in the world.[4] If the reader assumes for a moment that there is such a scientific discipline, never mind what we want to call it, then it is clear that such a discipline cannot be a matter of interpreting what Kant, Nelson or any other philosopher said or meant, in the same way that physics or biology are not about the right interpretation of Newton or Darwin.

What I have just described covers mainly what we may call the constructive part of Nelson's philosophical work—his ethics, his political philosophy, his philosophy of mathematics, and so on. History has not been kind to his particular constructions, and this is not the place to talk about or take a position towards them. The book in front of you belongs rather to the destructive part of Nelson's philosophy—his criticism of other philosophers. Some aspects of the method by which he analyses and picks holes in other people's arguments certainly make use of Kantian concepts and distinctions, some of which are controversial to say the least.[5] I'll come back to those in the next section. However, the most important feature of Nelson's method is a relentless use of quite elementary logic.

Our author very often applied his formidable analytic powers to refute the fallacies and weak arguments of his contemporaries. In particular, he wrote three long treatises for that purpose. The first, *On the So-called Problem of Knowledge* (1908), was dedicated to the refutation of all contemporary theories of knowledge and in fact of the very idea of a theory of knowledge, an idea whose inconsistency he tried to prove. The second, *Law Without Justice* (1917b), attacked the philosophy of law of many famous professors and reduced the doctrine of national sovereignty to tatters, a doctrine that was used then as now to oppose a functioning world order and a global rule of law. The third and last one, *Phantoms* (1921), aimed at the annihilation of the rampant relativism, nihilism, and scepticism that had started in the 19th century and, as we know, has continued in recent years to thrive under the name of post-modernism.

The present book is the distillation of the many battles Nelson fought in those treatises as well as in many shorter publications. In all of them he is always trying to

---

[4]In Nelson's own day, psychology as a scientific discipline was still in its infancy, and the method that still predominated in it—even in the lab—was some form of introspection. The experimental varieties of psychology (behaviorism, psychophysiology, neuropsychology, social psychology, cognitive psychology, developmental psychology) which use little or no introspection were at the time either very young or mere promises on the horizon. In consequence, Nelson did never go beyond a relatively naïve introspectionism. Thus to take seriously his idea of a science studying the conditions of human knowledge and action necessarily means going beyond him.

[5]Indeed, Nelson was led to his theory of philosophical fallacies by a careful re-consideration of Kant's famous, much-discussed and often rejected distinction between analytic and synthetic judgments. In fact, the equivocation fallacy Nelson identified as pervasive in the history of philosophy was explicated by him as starting with an analytic judgment and then surreptitiously putting a synthetic judgment in the place of it. In some cases, this replacement might have been a deliberate trick, but for the most part Nelson rather believed that philosophers were unconsciously self-deceived because of elementary logical and linguistic mistakes (in that respect, he was a kind of German analytic philosopher *avant la lettre;* see Berger et al. 2011).

show that philosophers, or nonphilosophers doing philosophy, will fall into equivocation and become entangled in false dilemmas whenever they do not follow the right method.[6]

Some Objections to the Theory

For reasons the reader will presently appreciate, we should stop for a moment to consider the meanings that the word 'fallacy' has in common parlance. There are basically three of them.

(1) *A fallacy is an argument that is logically invalid*. In this sense, people often talk of 'formal' fallacies—fallacies that go against the strictures of formal logic.

Otherwise we may talk of 'informal' fallacies, i.e. arguments that may be formally valid yet turn out to be offensive in other ways. Of these we have two varieties:

(2) *A fallacy is an argument at least one of whose premises is false*. If an argument is formally valid and all its premises are true, then that argument is sometimes called 'sound'.

The second sense of 'fallacy' and the first variety of 'informal fallacy' thus concentrates on the nonlogical aspect of 'soundness'—the truth of premises—quite apart from the question of the formal validity of the argument itself. This is perhaps the meaning of 'fallacy' most widely appealed to, although from the standpoint of pure formal logic the truth of the premises is perfectly irrelevant to the question of the validity of an argument. Usually, though, it is not any old premise which provokes an adversary to cry 'fallacy!'—but only premises that appear to have certain properties, e.g. the fact that they possess some methodological import or that they are somehow hidden or taken for granted. Now consider the case of an argument that has no such offending premise. A third sense of 'fallacy' and a second variety of 'informal fallacy' thus emerges:

(3) *A fallacy is an argument that evades its task of engaging with what the other person is saying*. In this case an argument may be formally valid and all its premises may be unimpeachable, and yet the objection is that whoever is presenting it is maneuvering the whole discussion far away from the point of interest. This is for instance the case when the conclusion repeats one of the

---

[6]Nelson had a positive theory—a theory of correct argumentation—to accompany the purely negative theory of philosophical fallacies contained in this book. Nelson first presented that positive theory in his doctoral dissertation (Nelson 1904); but perhaps the most complete exposition for the special case of ethics is Nelson's *Ethical Methodology* (1915). By the way, that theory is hinted at in Chapters "Lecture XII", "Lecture XIII" and "Lecture XVIII" in this book.

premises or when there is a subtle change of subject, so that we go off at a tangent.

Nelson's notion of fallacy partakes of the three meanings. The one big and typical philosophical fallacy he is at pains to identify, describe and illustrate in the present book is certainly a logically invalid argument, for it is a fallacy of equivocation. In order for any argument to be formally valid, the words we use in it must have a constant meaning, otherwise we will not be able to say that we are talking about the same things all along. So it is a fallacy in sense (1). Nevertheless, the 'philosophical fallacy' also uses at least one important false premise. That premise may consist in a questionable tacit assumption, very often a presupposition shared by two philosophers opposing each other. Or it may consist in a methodological legerdemain, consisting in the replacement of a synthetic by an analytic judgment, or the use of a definition (which as such cannot be either true or false) masquerading as a proper proposition, for it appears to be secretly endowed with the aura of truth that comes from the fact that a definition is in itself an arbitrary convention. It would then also be a fallacy in sense (2). Finally, by the use of dubious premises of the sort described, the argument will either be tautological or else immobilise the discussion by placing a questionable assumption out of the reach of critical discussion. And so it would be a fallacy in sense (3) as well.

Let us now examine the scope and limits of Nelson's theory. As far as I can tell, there are three main arguments, of unequal merit, which one can use against it.

Objection 1 If we concede for the sake of argument that the fallacy identified and profusely illustrated in the present book is actually a fallacy, and even a very frequent, indeed typical one, we could still argue that it is only *one* among many other fallacies committed by philosophers and described by other authors. There would be two ways to develop and meet this objection.

On the one hand, there is a thousand-year-old tradition of identification of fallacies, both formal and informal, started by Aristotle in his *Sophistical Refutations* and going all the way down to our own day. Although the study of fallacies is, to say the least, a very controversial subject (see Hamblin 1970), the idea that they are many and not just one seems to enjoy a pretty broad agreement.[7] Nelson himself does certainly not argue for a complete reduction of all fallacies to the one he pays so much attention to. First of all, he only talks of fallacies in philosophy, not of fallacies in general. Secondly, he concentrates on what is typical, i.e. he does not go so far as to say that there could not be other fallacies in philosophy, but only that the one he describes is by far the most frequent in philosophical argumentation. In statistical parlance, he has pointed out the *modal* fallacy in philosophy.

---

[7]Lawrence H. Powers (1995) has argued that all fallacies can be reduced to equivocation. His argument started from a consideration of philosophical method (Powers 1986) and led him all the way to the claim that not only all philosophical fallacies but also all nonphilosophical fallacies as well are cases of equivocation. His theory has some points of contact with Nelson's, although I unfortunately cannot here enter into a proper comparison of the two theories. I thank the two anonymous referees of this book for pointing Powers' work out to me.

Some Objections to the Theory 11

On the other hand, philosophers are well-known for the habit of inventing names for arguments that they reject as fallacious. Thus, just to quote two famous British contemporaries of Nelson, George Edward Moore (1903) declared that almost all writers on ethics committed the 'naturalistic fallacy' and Alfred North Whitehead (1925) introduced 'the fallacy of misplaced concreteness'. First of all, these and countless other so-called philosophical fallacies suffer from one disease—there is no consensus as to whether they are indeed fallacies, in fact there is no consensus as to what exactly they are. But secondly and more importantly, it *can* in fact be argued that in both cases the culprit is indeed the surreptitious replacement of a given concept ('good' in the case of Moore, any abstract noun in the case of Whitehead) by a made-up one, so they would be covered by Nelson's theory. It would be interesting to see whether other fallacious argument schemes proposed by philosophers can be reduced to Nelson's—a task that belongs to any serious critical reading of the present book.

Objection 2 All the examples of philosophical arguments analyzed by Nelson look like traditional syllogisms, whose premises and conclusion have a subject-predicate structure. Now if there is a lesson coming from the field of mathematical logic, it is that most arguments are not syllogisms and the logical form of most sentences used in argumentation cannot be reduced to that structure. This would suggest that Nelson's theory is very far from being applicable to *all* philosophical arguments. His is a logically outdated theory which should be replaced by something more in tune with modern logical insight.

This is undoubtedly a powerful argument—in fact devastating if it could be proved that the basic structure revealed by Nelson's diagrams necessarily preclude the use of logical forms different from those of traditional syllogistic logic. Yet if you consider the bare skeleton provided in the general diagram above (see p. 4, in this chapter), you will see that the propositions P and Q could have any logical form whatsoever. In fact, even a cursory inspection of many examples used by Nelson in this book reveals the presence of relational terms, which cannot logically be reduced to subject-predicate structure. Chapter "Lecture XIV" thus refers to the Austrian philosopher Franz Brentano's definition of 'good' in terms of 'love', which is a binary relation between something that loves and something that is loved. In fact, Brentano's definition is obviously, from a logical standpoint, a higher-order proposition in that the goodness of an object requires all (relational) 'events' or 'acts' of loving to be 'right'. Nelson, however, does not have to go into the complicated logical details of Brentano's definition in order to criticise it. His criticism of this and other pseudo-definitions (and arguments based on such pseudo-definitions) needs very modest logical means. The interesting question is then rather whether philosophical fallacies—the 'typical' ones—are all in fact rather elementary. This is indeed Nelson's tenet, and if any reader thinks we need much greater logical sophistication to fathom other profound fallacies in philosophy, let them try and show it. Till then we may be healthily sceptical of any such claims.[8]

Objection 3 Nelson's theory assumes Kant's two most famous distinctions of kinds of judgment—analytic versus synthetic and a priori versus a posteriori. These two distinctions were combined by Kant to produce the mixed class of synthetic a priori judgments—the absolute presuppositions of all human knowledge and all moral action. This whole system of thought has been controversial from the start. We should nevertheless distinguish two kinds of debate. Some philosophers after Kant have adopted some formulation of his distinctions yet tried to carve a distinctive non-Kantian position (e.g. the logical positivists or logical empiricists). Those philosophers might want to subject Nelson's particular formulation to some criticism yet would not reject his theory of philosophical argumentation only because it is based on Kant's distinctions. Other philosophers, however, foremost among them Quine (1951), have actually rejected the Kantian distinctions themselves as insufficiently clear and favoured instead a completely different framework in which to think about human knowledge or moral action.

If Quine's argument against the analytic-synthetic and the a priori-a posteriori distinctions had won the universal consensus in philosophy, then this objection would put serious doubts on the robustness of Nelson's theory.[9] For a while Quine's formidable reputation in the circles of analytic philosophy was such that an unusually wide consensus could not be denied, even though there were always dissenting voices. Nowadays that consensus has weakened considerably and there is a debate not only about the content of Quine's argument but in fact also about whether it *is* an argument (see e.g. Gutting 2009, pp. 11–30). Technical work in the mathematics of modal logic has made talk about the Kantian distinctions respectable again, and a considerable part of contemporary metaphysics, epistemology and moral philosophy is unthinkable without those distinctions.

So, this is what I would reply to the strongest objections to Nelson's theory I can think of. I do not flatter myself that my replies have destroyed the objections, but I hope at the very least they are sufficient for the reader to keep an open mind. After all, here as in many other fields the proof of the pudding is in the eating. Whether the theory of philosophical fallacy presented in this book is as powerful as its author believed it to be can only be decided by whoever takes the time to consider it fairly.

---

[8]To illustrate my point let us take the famous, and ingenious, claim by Geach (1958) that none other than Aristotle had committed a formal logical fallacy in the transition from 'For all human activity there is something for whose sake that activity is done' to 'There is at least one thing for whose sake everything is done' (*Nicomachean Ethics* 1094^{a}18–22). To talk shop for a moment, this would be a clear-cut case of an inference—an invalid one—from $\forall\exists$ to $\exists\forall$. That Geach's claim is highly dubious has been shown by Jacobi (1979), and besides it could be argued that, if there *were* a fallacy in the said passage, it would fall under Nelson's theory, it being a case of replacing the given concept of 'good' by an invented one, viz 'whatever it is that everybody wants'.

[9]By the way, the fact that Kant's distinction was originally formulated by him in terms of the subject-predicate structure is irrelevant. Even Quine (1951) does not consider this fact as important.

Note on the Translation

Nelson's original text carries the title *Typische Denkfehler in der Philosophie*, literally "Typical Errors of Thinking in Philosophy". For the reasons indicated in the Introduction, I think the best translation of *Denkfehler* in the sense Nelson uses the word is 'fallacy'. The title chosen for the English publication goes far beyond that word in order to make clear that Nelson is proposing no less than a *theory* of fallacious argumentation in philosophy (see the Introduction for more details). Other unorthodox translations in the text are justified in the notes.

The present book was originally a *Vorlesung*, that is a university course exclusively consisting of a series of one-hour public lectures, read aloud without interruption two days a week, Wednesdays and Fridays, during the Summer Term of 1921, which ran from April to July. The complicated political and economic situation in Germany at the time together with Nelson's engagement in multiple political and educational activities prevented him from following the schedule of an ordinary professor. The following table gives the details:

Wednesday	Friday
	29 April—Lecture I
4 May—Lecture II	6 May—Lecture III
11 May—Lecture IV	13 May—Lecture V
18 May—no lecture	*20 May—no lecture*
25 May—no lecture	*27 May—no lecture*
1 June—no lecture	*3 June—no lecture*
8 June—Lecture VI	10 June—Lecture VII
15 June—Lecture VIII	17 June—Lecture IX
22 June—Lecture X	24 June—Lecture XI
29 June—Lecture XII	1 July—Lecture XIII
6 July—no lecture	8 July—Lecture XIV
13 July—Lecture XV	15 July—Lecture XVI
20 July—Lecture XVII	22 July—Lectures XVIII and XIX
27 July—Lecture XX	29 July—Lectures XXI and XXII

It can be seen from this that Nelson had to pack two lectures into the last two Fridays of the Summer Term in order to fulfill his plan (even so one as the impression that he was rushing things a bit at the end). Incidentally, both the original typescript and the German edition separates the twenty-two lectures by indicating the date in which it was given. In this book I have replaced the dates throughout with Roman numbers. To help readers I have added a short abstract to each lecture, this being more informative than an arbitrary title.

This edition also differs typographically from the German one in a few respects. Throughout the book, Nelson is concerned with concepts in such a way that highlighting them by means of SMALL CAPITALS helps the reader to identify the targets of each discussion. I may have sinned by excess in this respect, in which case I hope my zeal will be forgiven. Whilst confessing my sins, I must also admit that now and then I tried to make a dense text easier to understand by using either *italics* (for emphasis) or inverted commas (to mark semantic ascent).

Again, whenever it seemed apropriate, I used the technique, originating with mathematicians and then adopted by linguists and philosophers, of numbering the sentences the author is focusing on. Very occasionally I also used a similar convention and added a few letters to denote variables as a means to make tracking reference easier. Anybody who has ever taught will know that Nelson must have written some words and sentences on the blackboard for everyone to see and follow the discussion; and indeed, there are hints in Nelson's text that he actually did so, e.g. on pp. 62, 66–67, 91, 140 and 156–157 (in this book). Numbering the sentences has a similar effect on the readers' attention.

Although I had to resist the temptation to change Nelson's diagrams to make them even more similar to the modern usages, arrows *were* added to indicate the direction of the reasoning involved, as is now usual in argument mapping. All in all, there are seven figures in the text. Six of those are themselves contained in Nelson's typescript. Figure 1 in Chapter "Lecture VIII" is a table displaying the four combinations of the analytic-synthetic and a priori-a posteriori distinctions. Figure 2 in Chapter "Lecture VIII", Fig. 1 in Chapter "Lecture X", Fig. 1 in Chapter "Lecture XIX", Fig. 1 in Chapter "Lecture XXI" and Fig. 1 in Chapter "Lecture XXII" are diagrams devised by Nelson to illustrate the dialectical state of affairs which he is discussing in each passage. From passages on pp. 83 and 91 (in this book) it can be inferred that Nelson himself drew those figures on the blackboard as he spoke (albeit without the arrows). He must also have drawn a figure when explaining the geometrical example on p. 122 (in this book), but no such drawing is extant in the typescript; so, after consulting a specialist in the teaching of mathematics, I added Fig. 1 in Chapter "Lecture XIII" in order to help make Nelson's point clearer to the reader.

Sometimes I felt the need to add a word, phrase or sentence to Nelson's text in order to make explicit what he meant. This additional material not in the original German typescript I have always enclosed within square brackets, following a common usage in quotations.

The typographical changes I have just described are all, as it were, of a logical nature; but there are three stylistic changes that deserve separate mention. One is the division into paragraphs, which I strove to adapt to modern English usage. Another is the introduction of appropriate indentations. Finally, pronoun usage was made to follow the rule of the 'generic masculine gender'. Any attempt at making the text politically correct in that respect is bound to be artificial in an author writing in 1921. On one occasion I even dared to translate German *Mensch* (that has the masculine gender, although it is supposed to be gender-neutral) for 'man' to keep it in line with contemporary English writings. I hope not to have offended anyone with this. In any case, the reader might be secure in the knowledge that Nelson was quite advanced for his time—men as well as women collaborated together in his circle on an equal footing.

Note on the Translation

A different kind of departure from the German edition is the inclusion, as an Appendix, of a long passage taken from another work (Nelson 1914), which I translated for the purpose of preventing a possible misunderstanding. In Nelson's lectures Kant appears throughout almost like an infallible philosophical hero, incapable of committing the fallacies identified, described and explained in the main text. Kant's role is always depicted as the discoverer of those logical insights upon which the analytic method deployed in this book rests. It is the purpose of the Appendix to show that, although a Kantian, Nelson was by no means dogmatic and quite capable of detecting and analysing those instances in which even the sage from Königsberg indulged in an occasional nod while thinking and writing.

Although I always looked for available English translations of all German texts (including German translations of texts in other languages) which Nelson quotes, in the end they were all freshly translated, including the two cases of humorous verse. In doing so italics, inverted commas and other idiosyncrasies of the original authors were respected, with one exception. The only alteration I introduced was the already-mentioned use of SMALL CAPITALS to highlight the concepts the author wants the reader to concentrate on.

The footnotes to the text in the Appendix stem all from Nelson, but I am alone responsible for those to the main text. The latter pursue three aims. The first and foremost is to make explicit Nelson's bibliographic references. My work in locating those was greatly facilitated by my predecessors—Erna Blencke and Friedrich Knigge for the German typescript, Andreas Brandt and Jörg Schroth for the German edition. All their attributions were checked and only a few were amended or added. Of course, the tracking down of English translations of German books or papers had to be done for the first time. To facilitate the scholarly reader's task of looking up the passages quoted by Nelson, I used either page numbers or authors' internal divisions (chapter, section, paragraph, and so on). This is especially important in the case of classical authors (like Locke, Berkeley or Hume), where there are so many editions available. The only exception is in the Appendix, where I followed exactly Nelon's method of reference. No harm was done, since the copyright-free translation I used is available on the net and thus easily accessible to readers.

The second aim of my footnotes is to contribute to a better understanding of the author's arguments by pointing to some contemporary or later contributions in philosophy, logic, mathematics or science generally. This has led me to add some new references, which will hopefully prove useful to Nelson's new readers.

The third and final aim of my footnotes is to express and confront possible objections to Nelson's arguments that may occur to somebody reading it from a current philosophical perspective. Although the Introduction contains my replies to eventual objections at a general level, there are some of a more particular character that had to be handled in situ, as it were.

One last terminological matter, brought to my attention by one of the anonymous referees. There are some important terms in Kant's philosophy and adopted by Nelson which might cause consternation. Among the most important are *analytisch*, *synthetisch*, *Anschauung*, *Schein*, *Pflicht*. As far as the translation is concerned, there should be no problem: they have customarily been rendered in English by 'analytic', 'synthetic', 'intuition', 'illusion' and 'duty', respectively; and I have found no reason to deviate from custom. As for the meaning of each of these terms, this is a question that has occasioned many controversies and will probably continue to do so. Although it would be absurd to pretend to resolve them here, some remarks might help readers to find their bearings.

Leonard Nelson thought of the Critique of Reason not as the name of a book or a series of books written by Kant, but as the name of a *discipline*, and an empirical one at that—a natural science of the human mind, whose mission it was to find out what the normative content of that mind is. So the said terms are meant to refer to *real* contents, whose existence was in part empirically proved by Kant; yet only in part, since Kant, being human, naturally made mistakes. To correct these mistakes and give the world an improved version of critical philosophy was Nelson's ambition, which he inherited from Jakob Friedrich Fries (1773–1843) and *his* followers. Nelson did not deceive himself about the fact that he must also have made mistakes; and he only hoped that later thinkers would make their business to correct them.

Analytic and synthetic judgments are thus, for Nelson, real products of the activity of our minds; intuition is a real capacity of our minds with real, tangible products (viz sense perception and mathematics); "illusion" is the name of mistakes to which we really and actually succumb every time we misuse or misapply concepts; and "duty" is the name of a principle which, whether we like it or not, whether we respect it or not, whether we rebel against it or not, really and actually guides both our actions and our thoughts about those actions.

Nelson broadly accepted Kant's doctrine about the illusion, intuition and the analytic-synthetic distinction, as it is set forth in the *Critique of Pure Reason*; the changes he suggested here or in his other works are minor. As for the concept of duty, Nelson believed Kant had fared less well. To give an idea of the mistakes he believed to have spotted, I have translated a few pages from a historical work on ethics (Nelson 1914) and placed them at the end of the lectures, as an Appendix to them. There may be some readers, however, who want to know the full results of Nelson's research in moral philosopy. I must refer them to his three series of *Lectures on the Foundations of Ethics* (Nelson 1971–1977, vols. 4–6), and especially to the first series, the *Critique of Practical Reason* (ibid., vol. 4), so called because, as I said before, Nelson conceived such a title not as a book by Kant but as a discipline to cultivate in the same sense as any other empirical science should be constantly extended and improved.

Acknowledgments This book in a way started life in 1983, when I spotted its contents in the shape of an old German typescript lying around in the bookshelves of the eminent psychologist and ergonomist Paul Branton in his London home. So my thanks should first go to my good old friend, who died too soon to celebrate this publication. In fact, as far as this book is my work, I would like to dedicate it to his memory.

I was very excited by my discovery and started pestering him with the sort of inquiries only an overeager 28-year-old can make. Was there an original manuscript? Why was it not included in its author's then recently published collected works (Nelson 1971–1977)? Were there any other transcriptions of Nelson's lectures? The answers to those questions will be given presently.

Many years later, Paul's widow and a great friend of mine, Rene Saran, asked me for advice as to which text by Nelson the Society for the Furtherance of the Critical Philosophy (SFCP) could endeavour to have translated into English for the purpose of making him better known in a world which in the meantime had thoroughly converted to English as the only global *lingua franca*. My immediate reaction was to point to the said typescript. I wrote to Dr Jörg Schroth, the young scholar who knew the Nelson archives papers best, and asked him whether he had stumbled upon the original MS of this book. He had not but was intrigued by my inquiry, and, lo and behold, he found it! Or at least, he found a pile of sheets, ostensibly typewritten by Nelson himself, yet altered in two ways. On the one hand it contained quite a few amendments, mostly grammatical corrections, written by at least two unknown hands. On the other hand it had a couple of paper strips pasted with full quotations of authors discussed in Nelson's course. These alterations conform to the practice within Nelson's circle of having his collaborators suggest improvements that would make the text clearer to nonphilosophers as a step towards eventual publication.

The typescript also had a handwritten note by Erna Blencke explaining that the task of completing the bibliographic references would be too difficult, so that it was agreed by the editors of Nelson's collected works in German not to publish the course. The reader must understand that Nelson's typescript was written in full in order to allow him to read it aloud according to the German tradition of the *Vorlesung* or course of lectures. Naturally, the original text referred to many authors but it contained very few exact references to the books and papers commented upon.

The next task was to prepare a proper critical edition on the basis of the two extant versions as well as to complete the bibliographic references. The SFCP charged Dr Schroth with both tasks. On the basis of that (unpublished) work, he himself, together with Professor Andreas Brandt, eventually secured a German publisher for the German text, the prestigious house of Felix Meiner in Hamburg. The book came out in 2011 as volume 623 of the series *Philosophische Bibliothek* ('Philosophical Library'), well known to all German students of philosophy. This by the way was preceded by another of Nelson's courses, on the philosophy of nature, which Dr Schroth, together with Professor Kay Hermann, had published in 2004.

In the meantime the SFCP was trying through me to locate somebody who would undertake the very difficult task of translating the book into English. This proved to be very difficult until Professor Dieter Birnbacher at the University of Düsseldorf suggested Mr David Carus, a fully bilingual doctoral student who had already collaborated in the recent translation of Arthur Schopenhauer's *The World as Will and Representation*. After a few months of hard work on the critical text kindly provided by Dr Schroth, David produced the first draft. It was now my responsibility to read it through and to make the necessary corrections both as to form (bringing Nelson's German closer to modern idiomatic English philosophical prose) and content (getting Nelson's points and arguments exactly right every time, something that required carefully annotating the text).

In order to make sure that the philosophical content of the translation came through to the casual reader, I asked two of my best former students, Christian Cervantes and Joaquín Galindo, both perfectly proficient in English, to read the second polished draft and to identify any obscurities and *non sequiturs* that might still remain in the text. With their able help I then produced a third draft, which was sent to an associate of SFCP and an experienced translator, Dr Celia Hawkesworth, to carry out a last reading of the third draft in order to detect any purely grammatical or lexical infelicities that might have escaped a non-native speaker such as myself. She did a magnificent job, saving me from many howlers and helping to produce a much more readable and correct text. As for the Introduction, my old friend Pat Shipley as well as, again, Rene Saran and Joaquín Galindo, went through it with a fine comb. The positive comments of Joaquín, who is a formidable logician, were a great relief to me, for when he says 'It does not follow', it usually does not. I owe Fig. 1 in Chapter "Lecture XIII", added to Nelson's text, to another former student, Marisol Radillo, whom I want to thank.

So I am deeply in debt with all these fine people, who have, each in their own way, helped me realise the longstanding dream to bring to the English public this *Theory of Philosophical*

Fallacies, born in 1921 yet capable (if I am not too far wrong) to bear fruit almost a century later. The SFCP deserves a separate acknowledgment, for it—or rather its wonderfully supporting trustees—not only financed the unpublished critical edition of the typescript as well as the publication of the German text, but also David Carus' first-draft translation, without all of which it would have been an even harder task to try and make Nelson speak to the English-proficient reader of the twenty-first century.

Two anonymous referees selected by the editors of this series have produced extremely interesting suggestions and corrections at all levels. These I have, to the extent of my abilities, endeavoured to learn from and to use for improving the final version. They did their best to point out my mistakes and I want to express my heartfelt appreciation for their help. All errors of fact, logic or style still remaining in any part of this book are naturally my sole responsibility.

Last but not least, I would like to thank my dear wife, Judith Suro, who lovingly coaxed me again and again to go back to the writing desk and finish what I had promised, and who always had a word of encouragement when I felt the whole thing was too much in the middle of my other academic obligations. As for my two children, Carolina and Rodolfo, I can only extend to them my sincerest apologies for all those occasions when they couldn't get through to an absentminded, Nelson-obsessed father.

References

Baggini, Julian, and Peter S. Fosl. 2010. *The philosopher's toolkit: A compendium of philosophical concepts and methods*. New York: Wiley.
Berger, Armin, Gisela Raupach-Strey, and Jörg Schroth eds. 2011. *Leonard Nelson—Ein früher Denker der Analytischen Philosophie?* [*Leonard Nelson—An Early Analytic Philosopher?*]. Berlin: Lit Verlag.
Bouveresse, Jacques. 1996. *La demande philosophique: Que veut la philosophie et que peut-on vouloir d'elle?* Paris: Minuit.
Cohen, Daniel. 2004. *Arguments and metaphors in philosophy*. Lanham (MD): University Press of America.
Cornman, James W., Keith Lehrer, and George S. Pappas. 1992. *Philosophical problems and arguments*, 4th ed. Indianapolis: Hackett.
Eemeren, Frans H. van, and Houtlosser, Peter. 2007. Kinship: The relationship between Johnstone's ideas about philosophical argument and the pragma-dialectical theory of argumentation. *Philosophy and Rhetoric* 40(1):51–70.
Geach, Peter T. 1958. History of a fallacy. *Journal of the Philosophical Association of Bombay* 5:19–20. [Reprinted in *Logic matters*. Oxford: Blackwell (1972), pp. 1–13].
Glymour, Clark. 1992. *Thinking things through: An introduction to philosophical issues and achievements*. Cambridge (MA): The MIT Press.
Gutting, Gary. 2009. *What philosophers know: Case studies in recent analytic philosophy*. New York: Cambridge University Press.
Hamblin, Charles L. 1970. *Fallacies*. London: Methuen.
Jacobi, Klaus. 1979. Aristoteles' Einführung des Begriffs 'Eudaimonia' im I. Buch der 'Nikomachischen Ethik': Eine Antwort auf einige neuere Inkonsistenzkritiken. *Philosophisches Jahrbuch* 86:300–325.
Johnstone Jr, Henry W. 1959. *Philosophy and argument*. State College (PA): Pennsylvania State University Press.
Johnstone Jr, Henry W. 1978. *Validity and rhetoric in philosophical argument: An outlook in transition*. University Park (PA): The Dialogue Press of Man and World.
Leal, Fernando. 2013. Leading a philosophical life in dark times: The case of Leonard Nelson and his followers. In *Philosophy as a way of life—Ancients and moderns: Essays in honor of Pierre*

References

Hadot, eds. Michael Chase, Stephen R. L. Clark, and Michael McGhee, 184–209. Malden (MA): Wiley.

Martinich, Aloys P. 2005. *Philosophical writing: An introduction*, 3rd ed. New York: Wiley.

Moore, George E. 1903. *Principia Ethica*. Cambridge (UK): Cambridge University Press.

Morton, Adam. 2004. *Philosophy in practice: An introduction to the main questions*, 2nd ed. New York: Wiley.

Nelson, Leonard. 1904. Die kritische Methode und das Verhältnis der Psychologie zur Philosophie: Ein Kapitel aus der Methodenlehre. *Abhandlungen der Fries'schen Schule (N.F.)* 1(1):1–88. [Reprinted in Nelson (1971–1977), vol. I, pp. 9–78. English translation: The Critical Method and the Relation of Philosophy to Psychology: An Essay in Methodology, in *Socratic Method and Critical Philosophy: Selected Essays* (pp. 105–157), New Haven, Yale University Press, 1949].

Nelson, Leonard. 1908. Über das sogenannte Erkenntnisproblem [On the so-called problem of knowledge]. *Abhandlungen der Fries'schen Schule (N.F.)* 2(4):413–818. [Reprinted in Nelson (1971–1977), vol. II, pp. 59–393].

Nelson, Leonard. 1914. Die kritische Ethik bei Kant, Schiller und Fries: eine Revision ihrer Prinzipien [Critical ethics in the works of Kant, Schiller, and Fries: A revision of principles]. *Abhandlungen der Fries'schen Schule (N.F.)* 4(3):483–691. [Reprinted in Nelson (1971–1977), vol. VIII, pp. 27–192].

Nelson, Leonard. 1915. *Ethische Methodenlehre* [*Ethical methodology*]. Leipzig: Veit & Comp. [Later published as part I of Nelson (1917b). Reprinted in Nelson (1971–1977), vol. IV, pp. 4–72].

Nelson, Leonard. 1917. *Die Rechtswissenschaft ohne Recht: kritische Betrachtungen über die Grundlagen des Staats- und Völkerrechts, insbesondere über die Lehre von der Souveranität* [*Law without justice: critical reflections on the foundations of public and international law, in particular on the theory of sovereignty*]. Leipzig: Veit & Comp. [Reprinted in Nelson (1971–1977), vol. IX, pp. 123–324].

Nelson, Leonard.1917b. *Kritik der praktischen Vernunft*. Leipzig: Veit & Comp. [Reprinted in Nelson (1971–1977), vol. IV].

Nelson, Leonard. 1921. *Spuk—Einweihung in das Geheimnis der Wahrsagerkunst Oswald Spenglers, und sonnenklarer Beweis der Unwiderleglichkeit seiner Weissagungen, nebst Beiträgen zur Physiognomik des Zeitgeistes: Eine Pfingstgabe für alle Adepte des metaphysischen Schauens* [*Phantoms—Initiation to the secrets of Oswald Spengler's art of prophecy, plus some contributions to the physiognomy of the spirit of the times: A pentecostal gift for all fans of metaphysical intuition*]. Lepizig: Der neue Geist. [Reprinted in Nelson (1971–1977), vol. III, pp. 349–552].

Nelson, Leonard. 1971–1977. *Gesammelte Schriften*, 9 vols., eds. Paul Bernays, Willy Eichler, Arnold Gysin, Gustav Heckmann, Grete Henry-Hermann, Fritz von Hippel, Stephan Körner, Werner Kroebel, and Gerhard Weisser. Hamburg: Felix Meiner. [This edition contains all of Nelson's texts except for Nelson (2004) and Nelson (2011)].

Nelson, Leonard. 2004. *Kritische Naturphilosophie*, ed. Kay Herrmann and Jörg Schroth. Heidelberg: Winter.

Nelson, Leonard. 2011. *Typische Denkfehler in der Philosophie*, ed. Andreas Brandt and Jörg Schroth. Hamburg: Felix Meiner.

Popper, Karl. 1979. *Die beiden Grundprobleme der Erkenntnistheorie*. Tübingen: Mohr. [English edition: Popper, K. (2008). *The Two Fundamental Problems of the Theory of Knowledge*. London, Routledge. The original German text was written by Popper in the late 1920s or early 1930s].

Powers, Lawrence H. 1986. On philosophy and its history. *Philosophical Studies* 50(1): 1–38.

Powers, Lawrence H. 1995. The one fallacy theory. *Informal Logic* 17(2): 303–314.

Quine, Willard van Orman. 1951. Two dogmas of empiricism. *The Philosophical Review* 60(1):20–43.

Rescher, Nicholas. 2001. *Philosophical reasoning: A study in the methodology of philosophizing*. Oxford: Blackwell.

Rescher, Nicholas. 2006. *Philosophical dialectics: An essay on metaphilosophy*. Albany (NY): State University of New York Press.
Rescher, Nicholas. 2008. *Aporetics: Rational deliberation in the face of inconsistency*. Pittsburgh (PA): The University of Pittsburgh Press.
Rosenberg, Jay F. 1996. *The practice of philosophy*, 3rd ed. Upper Saddle River (NJ): Prentice-Hall.
Schnädelbach, Herbert. 2012. *Was Philosophen wissen—und was man von ihnen lernen kann*. Munich: Beck.
Stove, Daniel. 1991. *The plato cult and other philosophical follies*. Oxford: Blackwell.
Vaz Ferreira, Carlos. 1916. *Lógica viva [Living Logic]*. Montevideo: Barreiro y Ramos.
Whitehead, Alfred N. 1925. *Science and the modern world*. London: Macmillan.
Williamson, Timothy. 2007. *The philosophy of philosophy*. Oxford: Blackwell.

Lecture I

Abstract Philosophy, as the search for truth, is a matter of thinking, reasoning and arguing correctly; and so the interest in truth implies trying to avoid fallacies. Intuition cannot be a source of knowledge allowing us to attain truth in philosophy; in fact, the results of intuition-led philosophy contradict both the facts of experience and each other. Current fashionable forms of pseudo-philosophy are averse to reasoning, for they either despair of ever attaining truth or else trust in intuition as their guide. Nonetheless, a certain "feeling for truth", which is not the same as intuition, is crucial for philosophical thinking and argumentation.

The title of this series of lectures[1] reveals at once the conception of philosophy which underlies my take on the relevant issues. Anyone who speaks of typical fallacies in philosophy presupposes that his audience is as interested in avoiding them as he is himself. This interest is none other than the interest[2] in truth, and whoever, driven by that interest, inquires about typical philosophical fallacies presupposes that reasoning and argumentation play at the very least a particular role in philosophy, that they are an important means, if not the necessary and only one, to attain philosophical truth. Yet this view is not universally agreed upon these days, astounding as that may seem. Indeed, nowadays it is certainly not at all

---

[1] The original title was 'Typical Fallacies in Philosophy'. It was changed in this translation to highlight the fact that Nelson was actually trying to develop a *theory* of philosophical fallacies. He himself makes this intention explicit in his next lecture, and he confirms it at the end of the last one.

[2] The concept of INTEREST (*Interesse*) is of great importance in Nelson's philosophy (as well as in the work of many other German-speaking philosophers, e.g. Habermas). Interests have an objective as well as a subjective side to them. Objectively, whoever has a stake in something has *eo ipso* an interest in it. Subjectively, there is an affective as well as a volitive element in interest, no matter whether the interest is theoretical (cognitive) or practical (active). Among human theoretical interests, what Nelson will in Chapter "Lecture XIII" call 'logical' interests occupy a special place in these lectures, e.g. the interest in uncovering fallacies, the interest in truth, the interest in definitions, the interest in proof and provability, the interest in rigour, the interest in consistency, and so forth. Whenever the word *Interesse* can be literally translated without sinning too much against English usage, I retain it. In any case it is good to remember that the word 'interest' as a philosophical term was quite common in all languages of eighteenth-century Europe, although it kept a terminologically central place only within German philosophy.

self-evident that someone who aspires to teach philosophy is driven by an interest in the truth, or moreover that he or she intends to apply arguments in order to arrive at that truth. In fact, it would be true to say that nowadays the majority of those professionally involved in philosophy consider this view dated and obsolete. For this reason, and before I begin with my subject, I will have to justify why I intend to appeal to this interest in the truth and moreover why I invite you to accompany me along the path of reasoning and argumentation.[3]

Anyone who is somewhat acquainted with today's philosophical literature will not be overly surprised when I state that the interest in truth and hence the method of reasoning are not held in especially high regard. They will know that pursuing the truth is of little interest to those who today call themselves philosophers, and that they will not generally say that what they do is ultimately concerned with what is true. For quite some time now people interested in philosophy have been in the grip of doubts about truth. They point to the history of philosophy, to the innumerable, recurrent attempts to grasp what is true. These continued attempts have resulted in a vast wasteland of systems that battle against each other, yet where no single system arises victorious and is able to survive for any length of time. What can explain this if not the fact that it is a mistake from the outset to want to find out something about what is true in philosophy? People who use the word 'truth' usually do so in a derivative meaning. They say, to be more precise, that they just aspire to a relative truth, i.e. truth in the sense that anyone who uses the word believes that what he says is true yet only for himself or at best for the times in which he lives, without claiming his doctrines to be universal and binding for other people or other times.

And where you actually find philosophers who have not given up on truth, who are seriously trying to discover it, you find that they seldom apply methods of reasoning and argumentation or at least refuse to believe it might be a good idea to use those methods. They are privy to a seemingly higher faculty that directly grasps philosophical truth. They call it intuition. Of course, anyone who possesses such a faculty can grasp truth in philosophy immediately and without the strain of

---

[3]Nelson follows Kant's practice of distinguishing two fundamental kinds of cognitive activity—thought (*Denken*) and intuition (*Anschauung*). When human beings manage to combine thought with intuition, they are able to achieve knowledge (*Erkenntnis*), yet thought by itself, without any intuition, only creates pseudo-knowledge; or as Goya, a contemporary of Kant, put it, 'when reason dreams, it produces monsters'. In Kantian philosophy thought embraces both the three fields of pure traditional logic (concepts or terms, judgments or propositions, reasoning or inference) and the fields of traditional 'applied' logic or 'methodology' (the study of fallacies, heuristics and research methods, the construction of theoretical systems). Most of the time German *denken* (the verb as well as the noun) must be translated by 'reasoning' or even 'argumentation'; yet in a few passages 'conceiving' is a more appropriate translation. *Denkfehler*, literally 'error in thinking' is best translated by 'fallacy' as used in modern argumentation theory, i.e. as a label for errors beyond obvious violations of formal rules of logical inference. We avoid the usual terms 'deduction' and 'deduce' for ordinary logical inference (excluding induction, abduction, and analogy) because in the Kantian tradition the word *Deduktion* was reserved for a special kind of transcendental argument which, in the post-Kantian tradition initiated by Fries to which Nelson belongs, goes beyond formal logic as usually conceived.

reasoning. I do not want, at present, to subject this point of view to further criticism. I would rather we understand it first, and then take up a position on it. I would also like to give you a psychological explanation of why this state of mind—so unfavourable to the pursuit of truth or at least to reasoning—is common among people who cultivate philosophy today.

I see the first reason for this in the fact that after a long period of lack of interest in philosophy, of indifference with regard to the great philosophical questions, a certain reaction has taken hold, a reaction that is not driven for the most part by proper academic research, but rather by a *feeling* that rebels against the dominance of empiricism—the long-standing doctrine that experience, or the mere investigation of facts, is everything. In opposition to empiricism, we find feeling struggling to vindicate the rights of philosophy. People aspire to a position *above* the facts, a position that might allow them to be judge and master of the facts, i.e. a position that could rightly be called philosophical. Yet this *feeling*, finding expression on all sides, is easily confused with a faculty that is in itself sufficient to take hold of philosophical truth. The feeling people talk about apparently amounts to some form of intuition. People in fact speak of the intuitive apprehension of the truth and believe that the aforementioned feeling is an apprehension of that sort. Whoever relies exclusively on this feeling is in danger of mistaking it for a faculty that can itself take hold of philosophical truth. They think they can immediately grasp the essence of things. Indeed, to exercise philosophy on the strength of such a feeling leads to a purported apprehension, explanation, knowledge, or what not, of the essence of things, or of the last things, as the fashionable expression states.[4] Those last things, which really should be last for our limited human minds, come to be the starting-point, and the task of philosophy seems to be to derive the individual and finite (which should be the starting-point for the ordinary human mind) from the apprehension of the absolute; yet the singular and finite are in fact the first and most immediate things for the common human mind.

The naïve onlooker will be worried when he realises that the different philosophers who proceed in this way are in such disagreement. If they all possess a

---

[4]See for instance Weininger (1904). I remind the reader that Wittgenstein was an admirer of this celebrated thinker of *fin-de-siècle* Vienna. A similar intuitive philosopher of Nelson's times was Oswald Spengler. Nelson discusses him below in this lecture and again in Chapter "Lecture XXII". It may be useful to compare a passage from an earlier book: 'This *feeling* is a dark awareness of what is true. It is therefore misleading to call it an act of intuition, as is often done. People speak of the 'intuitive apprehension' of something that is true yet only mean a dark awareness which in itself lacks all intuitive clarity... Feeling is no intuition but an act of reflection, even though it is different from the grasping of concepts or the drawing of inferences. Thus we have occasionally the feeling that an argumentation we have listened to or read is fallacious and yet cannot quite say what the fallacy is. We are confident that we shall find it, though, once we can think about it in leisure.' See Nelson (1917, 304). For the correct understanding of the first three lectures as well as the last two ones it is very important that the reader keeps in mind that the term 'intuition' for Kant, Fries and Nelson always refers to our sensory capacities; this includes the 'formal' intuitions that underlie mathematical knowledge but excludes all *intellectual* intuition. It is only *that* last capacity which is criticised both here and elsewhere.

faculty that allows them to apprehend the truth in philosophy or even the absolute essence of things immediately, it must seem a strange thing that the absolute essence turns out to be something different for every philosopher and that not even those who purport to possess this higher faculty can agree amongst themselves about the object they claim to be concerned with. And, as is perhaps even more apparent to the common person, he realises that the results of this philosophy contradict the facts of experience, which are accessible even to him and his ordinary intellect. All of this causes us to be sceptical of the type of philosophy carried out by those who in today's world at least aim at the truth.

But precisely the difficulties they encounter, the contradictions in their results, not only amongst themselves, but also with experience per se, make it plain that their pretension to pursue truth cannot be maintained long while doing philosophy in this way. Such a pretension cannot stand in view of those contradictions. As a result the intuitive philosophers drop it and instead claim that the meaning of philosophy lies in grappling with what cannot be established as the truth, what is far deeper than the truth, the irrational, i.e. what lies beyond rational experience. They thus come to make demands on philosophy which are more of an aesthetic nature; philosophy is to be exercised as a fine art and is to be judged according to the artistic depth of its results. No wonder then that by discarding the regard for truth this undertaking becomes increasingly arbitrary and fantastical.

This supposedly artistic way of doing philosophy has one thing in common with appeals to the higher faculty of extrasensory intuition—an enmity towards reasoning, a hatred of rational thought. In older days philosophy did not respect the rights of experience, so that where it was shown to be inconsistent with the facts, its contempt for the simplistic approach of the empirically-minded person was expressed in the reply, 'So much the worse for facts!' When today's philosophy is accused of being unscientific, of contradicting itself and of being non-logical, the parallel reply is, 'So much the worse for logic!' Kant in his old age published a short paper criticising this fanciful form of doing philosophy, which was of course as rife at that time as it is today. He speaks in that paper 'of a newly-raised sophisticated tone in philosophy'.[5] The new sophisticated tone expresses itself in its aversion to rational thought, to reasoning.

There is a further side to the sophisticate approach. It is not merely contrary to reasoning, but also to what some people find disagreeable about reasoning, namely the *work* which accompanies it. The sophisticate philosophy is basically a *work-averse* philosophy. Proof of this can be found in the articles and books that have become so common nowadays and are praised as being the spirit of the times in the field of philosophy. Doing philosophy the sophisticate way is celebrated for not being a matter of work, but a matter of creation, and those who do philosophy in this way flatter themselves with the notion that they will be reckoned to be artists or even that they were born artists. For he who has not been born an artist is in the very

---

[5]This paper by Kant was published in 1796 and can now be read in English translation in Allison and Heath (2002, 431–445).

same position as the researcher—he has to work. Indeed the born artist (one might refer to history on this, or to the artist who will himself confirm it) has had to work in order to bring forth precisely those things which allow him, in the view of dilettanti, to appear as the born creator. Doing philosophy the sophisticate way is thus not able to deliver what it promises. It is in fact no more a true art than it is true science or true scholarship.[6] Work belongs to true art just as it does to true science and true scholarship. The opposite term that matters here is work not art. Art, science and scholarship are in this sense truly akin. There are artistic, scientific and scholarly dilettanti and the one feature that marks them all is their aversion to work. A dilettante is someone who applies himself to tasks which he does not have the strength to fulfill, for such a strength can only be developed step by step with continuous effort and diligence. The aversion to work, typical of sophisticate philosophers, explains why they are as far from real art as they are from real science or real scholarship. Hence, they break with both, for they can no more live up to the expectations of cultivated taste than they can to those of research.

There is ample evidence of this if we consult today's literature. The philosophical writer who is read today by more people than anybody else, Oswald Spengler, in a book of over 600 words claims not to communicate the results of his research but instead only to express something that, as he himself says quite literally, can be felt through wordless understanding (Spengler 1918; see Nelson 1921). This deep, wordless understanding, which by all accounts has made such a powerful impression on our contemporaries, is expressed over 600 pages of wordy text. The same peculiarity, yet in another form, can be seen when philosophers of this type tell their students they are to learn to read between the lines. We can only hope that this approach, which is quite contrary to the manner in which the results of research are studied, might become even more consistent with its aim. If we are in fact to find real wisdom between the lines, then it would be best to make the space between the lines as large as possible and in fact leave out the lines altogether. There can of course be no greater act of kindness towards the public, to whom the price of books has become quite unaffordable.

It is well known that in his time Kant fought this fanciful way of doing philosophy when it became prominent in the particular field of natural science from which we can today consider it expelled. At the time it came on stage under the name of *Naturphilosophie*, but was not successful for any significant period of time.[7] The natural sciences, at least the more rigorous ones, that were cultivated with mathematical methods, were too well established to fall victim to this type of philosophy for long. This type of philosophy, which termed itself *Naturphilosophie*, was soon recognised as dilettantism and forced out of the scientific field. It then spread into another field where it has become even more firmly established—the so-called cultural sciences. People speak today of *Kulturphilosophie* and pride

---

[6]My translation pairs science and scholarship in order to render the German word *Wissenschaft*, which covers both endeavours.

[7]On *Naturphilosophie* see Ostwald (1902, First Lecture) and Kuhn (1977, 97–100).

themselves in doing cultural not natural philosophy. They do not want to come into contact with the natural sciences, in fact they specifically avoid this, and seek to fulfil the ambitions of a true philosopher by keeping as far away from the scientific field field as possible. Hence, they prefer to speak of the philosophy of culture. But in the eyes of a thoughtful judge this situation does reveal something. It does not commend this way of doing philosophy, for anyone who contributes to culture, anyone for whom that word 'culture' is important, uses the word sparingly. The more the word 'culture' is used, the more certain a critical onlooker can be that a considerable amount of real culture is missing. Missing here is precisely what I previously termed the shortcoming of this way of doing philosophy—that it lacks all connection with work. Culture itself, namely, revolves around work and is always to be found wherever there is solid work. Culture is present wherever and whenever one serves a cause with devotion—and the more devoted to it, the less one is prone to name-dropping and wordiness. Now the cause a real philosopher is to serve with devotion, is (I think) the truth—the truth in every field, nature, culture, whatever.

If we were to ask those who philosophise today what they would like to replace truth with, what they actually seek, we would infer from what they themselves state, and through the observation of what they actually practise, that it is more or less what Hegel saw as the philosopher's task, when he said that philosophy is 'its time captured in thought'.[8]

These philosophers are concerned with the analysis of the spirit of the times. Anyone who desires to be considered a philosopher these days must, above all, get in touch with the spirit of his time. In order to do this he must first track down the spirit of his time: he must get acquainted with it, become its confidante, let his own spirit become a mirror of it (as in Goethe's *Faust*); and this mirroring of the spirit of the times is what is today passed off as philosophy.

Now this spirit of the times is a funny thing, as Goethe revealed in his *Faust* [Part One, verses 575–577]:

> The spirit of the times—what you call that,
> That is at bottom just *your* spirit,
> From which the times bounce off reflected.

The spirit of the times—this great unknown, which the philosopher ought to have an intimate relation with—is in fact, when you look closely, nothing other than the fashion the philosophical literati themselves dictate and then pass off as the spirit of the times. Perhaps you read the article in the *Frankfurt Times* a few days ago: 'Overcoming Relativism' by Professor Arthur Liebert, the president of the Kant Society. In this fairly extensive article there was not one mention of the question of the truth of relativism, and if you were to ask why relativism is to be overcome, or has been overcome, then it was not because of a scientific weakness of this doctrine, but rather because the spirit of the times was in the process of

---

[8] See Hegel (1821), Preface.

turning away from it. Anyone who cares to listen to the development of the spirit of the times will notice that the times are turning away from relativism and whoever wants to be just one step ahead of his contemporaries would best express this by abandoning relativism. That was the argument of this article. Another example: a book that has just been published, that has been received with much jubilation, indeed has caused a true furore among friends of philosophy, is *The Resurrection of Metaphysics* by Wust (1920). If you follow the line of argument in this book, and ask why metaphysics is to be resurrected, you will not in fact discover that the enmity towards the metaphysics of the past has resulted from an error, a mistake that has now been set right with good arguments or by showing that metaphysics is a legitimate endeavour. Instead, Wust argues that the 'cultural spirit' has again expressed a need for metaphysics; and whoever shares the needs of the cultural spirit must now turn his attention to the search for the new metaphysics. In short: the cultural spirit, that for such a long time had to make do without metaphysics, is now bored and wants some change. So now it's time to recall metaphysics from her exile and put her back on the throne. What does this all mean, a thoughtful person will ask, considering perhaps that the cultural spirit will sooner or later have its need for metaphysics assuaged and could again grow tired of her? Anyone who already lives in that future period will be even further ahead of his time. And the game will be repeated over and over. It is thus not possible to understand how one idea could be preferred over another, other than by the fact that it is *fashionable* at the time it is written and spoken about.

It is no surprise then that with this conception of the nature and task of philosophy, no advance can be made in philosophy from one period to the next, and that those who take philosophy seriously lose their drive when, even among philosophers, they only observe a fluctuation from one opposing view to the other, a disagreement without end, where the one tears down what the other had constructed.

I stated earlier that what has attracted the educated public back to philosophy is an intensely powerful feeling. This could be construed as though I was trying to express a view much like the one I just described, concerning the resurrection of metaphysics. That is not what I mean. The feeling I am talking about here has a deeper significance. It is a feeling for what is true, an intimation of truth. This feeling for the truth is something different from a feeling for the needs of the spirit of the times. Yet the question is: what is the importance of reasoning for philosophy? What is the relationship between reasoning and a truth that expresses itself in the form of feeling? It might be thought that feeling is enough, and thus that we do not require reasoning. This is based on the view that feeling can be taken to be a kind of intuition, which is all we are supposed to need to determine what is true in philosophy. In my next lecture I will tackle the question of what reasoning can contribute to the attainment of truth in philosophy. And this will then justify why we should be interested in avoiding the errors of reasoning, the fallacies that occur in philosophy.

References

Allison, Henry, and Peter Heath, ed., 2002. *Immanuel Kant: Theoretical philosophy after 1781*. New York: Cambridge University Press.
Hegel, Georg Friedrich Wilhelm. 1821. *Grundlinien der Philosophie des Rechts*. Berlin: Nicolai. [English translation: *Elements of the Philosophy of Right*, Cambridge University Press, 1991].
Kuhn, Thomas S. 1977. *The essential tension*. Chicago: University Press.
Nelson, Leonard. 1917. *Kritik der praktischen Vernunft* [*Critique of practical reason*]. Göttingen: Vandenhoeck & Ruprecht [Reprinted in Nelson (1971–1977), vol. IV].
Nelson, Leonard. 1921. *Spuk—Einweihung in das Geheimnis der Wahrsagerkunst Oswald Spenglers, und sonnenklarer Beweis der Unwiderleglichkeit seiner Weissagungen, nebst Beiträgen zur Physiognomik des Zeitgeistes: Eine Pfingstgabe für alle Adepte des metaphysischen Schauens* [Phantoms—Initiation to the secrets of Oswald Spengler's art of prophecy, plus some contributions to the physiognomy of the spirit of the times: A pentecostal gift for all fans of metaphysical intuition]. Lepizig: Der neue Geist [Reprinted in Nelson (1971–1977), vol. III, 349–552].
Nelson, Leonard. 1971–1977. *Gesammelte Schriften*. In 9 vols. eds. Paul Bernays, Willy Eichler, Arnold Gysin, Gustav Heckmann, Grete Henry-Hermann, Fritz von Hippel, Stephan Körner, Werner Kroebel, and Gerhard Weisser. Hamburg: Felix Meiner.
Ostwald, Wilhelm. 1902. *Vorlesungen über Naturphilosophie* [*Lectures on natural philosophy*]. Leipzig: Veit & Co.
Spengler, O. 1918. *Der Untergang des Abendlandes: Umrisse einer Morphologie der Geschichte*, vol. 1: *Gestalt und Wirklichkeit*. Munich: Beck [English translation: *The Decline of the West: Form and Actuality*, New York, Knopf, 1926].
Weininger, O. 1904. *Über die letzten Dinge*. Vienna: Braumüller [English translation: *On Last Things*, London, Edwin Mellen, 2001].
Wust, Peter. 1920. *Die Auferstehung der Metaphysik* [*The resurrection of metaphysics*]. Leipzig: Felix Meiner.

Lecture II

Abstract The struggle between our original feeling for truth and opposing interests creates a sophistic 'dialectic' in which there seem to be arguments for all sorts of incompatible philosophical statements. Given that philosophical truth is not intuitive, we can only defend our feeling for truth by engaging in sound philosophical argumentation. So these are lectures in logic as applied to philosophy; and what they strive for is a general theory of dialectical error, in which the typical and most frequent philosophical fallacies are dissected in the light of actually occurring examples.

In my last lecture I attempted to point out the reasons why people are these days ill-disposed towards a philosophy which considers reasoning its principal means of research and to identify the sources of this aversion to reasoning among philosophers. I do not want to delve into that again, but would like to draw on our last conclusion. We had discovered a particular reason for this popular view, namely that we can actually point to a specific, allegedly higher, faculty, which allows people to get by without reasoning. This faculty is what I called our feeling for the truth.

If we look more closely at what philosophers of intellectual intuition appeal to, we will in fact discover that they misinterpret the scope and power of that feeling for the truth in their work. It is therefore important to clarify what exactly this feeling for the truth is and investigate what its relationship is to reasoning. Let us assume for the sake of argument not only that there is such a feeling (after all, we all know it from experience), but also that there are good *prima facie* reasons to trust it. Let us then abide by the decisions made by this feeling for the truth. What kind of experiences do we then have? We observe that this feeling, which we like to trust and would like to rely on entirely, lets us down by getting embroiled in self-contradiction. The dicta of one person's feeling for the truth are incompatible with those of somebody else, nay even sometimes with each other and within the same person. If feeling seems to be a suitable means for judging philosophical truths, it also turns out to be subject to the danger of confusion and falsification. How else would it contradict itself? So we face the question of which of the two contradictory feelings we should trust, and which we are to view as a falsification, a distortion of the originally sound feeling. For an answer to this question we must look elsewhere. We have to seek it outside feeling and hence be prepared to reason.

If we consider more closely how such a falsification and distortion of the originally secure feeling for truth has come about, we will realise that in fact what we call dialectics, hence reasoning, plays a significant role.[1] If we set about philosophising we are no longer naive, but rather have already made use of our rational understanding; and this use has influenced our feeling. Thus we can no longer view our feeling as entirely unbiased and therefore as a reliable, pure source of truth. Before we begin to philosophise consciously and intentionally, we are all subject to the influences of a dialectic, which in any society we might live in unavoidably exerts control over people's minds and thus also over our own. The influences of this dialectic are what have prevented our feeling from speaking with one clear voice. What drives us to this dialectic is another question. I only wish to relate the fact that such a dialectic exists and that it influences our minds, whether we like it or not.

If we ask ourselves what it is that drives people to dialectic, it is mainly *interest*: desires, hopes, fears, that cause people to reflect, and hence to put into motion a mechanism which in its final results brings about a change in their judgment, a judgment that had originally been based on their feeling for the truth. In the first instance, we resist a judgment thus based on feeling because its conclusion is contrary to our desires and hopes. Hence, we begin to reason, to brood, to reflect and to replace the voice of that pure feeling with a distorted judgment based on interest, which results from sophistic processes of reasoning. In the end, it is difficult to decide what comes from such a sophistic sleight of hand and what from the voice of the pure feeling for the truth.[2]

However, it is not simply selfish interest that leads us to such dialectics, and therefore to a falsification of the original feeling for the truth. In fact, interest in the truth can play a similar role here. Interest in the truth can cause us to inquire into the reasons for the feeling-based decision. It awakens the desire in us to argue

---

[1] Nelson uses the word 'dialectics' or 'dialectic' (*Dialektik*) in the traditional sense, i.e. to indicate the back and forth of argumentation that takes place between different thinkers who are prepared to back up their positions with reasons.

[2] The phrase 'sleight of hand' translates *Erschleichung*, a term from German Law which in turn translates *obreptio* or *subreptio* in Roman Law—a *surreptitious* (same root) misrepresentation of facts to obtain a favour or service. A series of German philosophers from Leibniz to Kant borrowed it to denote a specious argument by which concepts are replaced or swapped for each other (see Birken-Bertsch 2006). More generally, traditional logicians used *subreptio* or *vitium subreptionis* to denote the verbal equivocation produced by using one term for two different concepts, a fallacy whose best-known manifestation is the *quaternio terminorum* in a classical syllogism. Starting with Chapter "Lecture IX" below Nelson will use *Erschleichung* (or more fully, *begriffliche Erschleichung*) to indicate a mixture of these two senses: the dialectical context in which the two parties in a philosophical debate reciprocally swap their concepts. At that point the phrase 'sleight of hand' becomes too imprecise a translation and it will be replaced by 'swapping of concepts' or 'concept-swapping' as the core of the philosophical fallacies Nelson is discussing.

conclusively in favour of those judgments that are initially only felt, and for this argumentation we need dialectics. If we are not able to establish a dialectically sound argumentation, then dialectics in this case also leads to a falsification of our original judgment.

Regardless of which interest might be decisive, there is still a sophistic dialectic whose influence cannot be avoided by anyone and which we are already subject to when we begin to philosophise methodically and intentionally. Our feeling of truth is powerless against this sophistic dialectic. We can resist it only by drawing on a *better dialectic*, in order to disarm the sophistic one. Of course, this requires effort and perseverance and it is easier and more tempting to oppose this sophistic dialectic with the notion of a purported intuition of philosophical truth. In this case, we replace feeling with intuition and declare the latter an adequate guarantee of truth. But it is clear that such a method is insufficient and unsatisfactory from the standpoint of proper scholarly inquiry. For the intuition people have recourse to is, when you look closely, precisely what is attacked by the sophistic dialectic we wish to resist. One is replying to the doubts and objections of such a dialectic by a sheer assertion when one appeals to a purported intuition, when this is precisely what is contested by the said dialectic. The fact that it is open to attack by such a dialectic is the best and most unmistakable proof of the fact that the purported intuition is purely fanciful. For if we really were in possession of such an immediate intuition of philosophical truth, then no sophistic dialectic could have any success whatsoever against the truth. If it can, and it is a fact that it can, then philosophical truth is incapable of revealing itself to our mind with immediate clarity and therefore intuitively. If we wish to secure this truth, then the sole means of achieving this is through *reasoning*.

Let us now look out for an explanation of the defects associated with the feeling for truth and for the reason why it needs to be supplemented with, and corroborated by, reasoning. If we consider this question, we will realise that there is a peculiar quality to this feeling for the truth—it only comes to the fore when we judge particulars. Only in the judgment of *this* case, which we are *here and now* confronted with, does our feeling for the truth speak out. However, it always fails as soon as we ask for the general reasons for such a judgment. As soon as this question arises things become opaque and we are faced with that struggle between contrary propositions which is so characteristic of the philosophical enterprise. For as long as we do not commit to philosophy, and that means, for as long as we remain concerned with particulars that life presents to us, and do not pose any *general* questions, our feeling for the truth guides us reliably and assuredly. The conflict begins when we generalise the question, that is, as soon as we ask for the *reasons* that we, in judging the particular case, seem to be quite sure of and indeed act on. Finding out the general reasons for judging the particular case requires reasoning, and this reasoning is what we properly call philosophising. We need dialectics, i.e. a method of abstract reasoning, as soon as we leave the judgment of particulars behind and look for the general principle.

The history of philosophy is concerned with seeking an increasingly satisfying solution to the task of securing the general principles for all possible judgments of particulars. For this reason, real progress in the history of philosophy takes place in the dialectical field. In judging specific facts philosophers do not disagree amongst themselves nor with the judgments of everyday human understanding. The disagreement concerns those general principles, and dialectics is needed to establish them.

I said that real progress in the history of philosophy was to be found in the dialectical field. We can state even more specifically: it is to be found in the field of *logic*. For this much is clear: if we have but once seen through the misconception of mysticism, of a purported higher intuition of the sources of philosophical truth, then we will also recognise the futility of hoping to expand the content of the general principles of our knowledge through philosophising. This content of the general philosophical principles is neither created nor expanded by reasoning. For the expansion of our knowledge, such as is achieved through experience, is not what we are concerned with when we speak of *philosophical* truth. Reasoning is of no use to us here other than to free those general principles, which we presuppose in every genuine feeling for the truth, from the obscurity with which they are originally conceived and to elevate them to the complete clarity of consciousness. It is only with regard to this varying degree of clarity that individual thinkers differ. This clarity is achieved through the scientific method of tracing back *all particular judgments to their general principles*, through the progressive analysis and clarification of the general presuppositions that a mere feeling for the truth is unable to grasp in the abstract. This *scientific method*, which is achieved through the reduction of particular judgments to their general principles, is none other than the method of *explicitly logical reasoning* in contrast to the implicit application of principles in judgments made through a feeling for truth. And the rules for explicitly logical reasoning in contrast with merely felt judgments, are none other than the *laws of logic*. Hence, the art of explicitly logical reasoning only develops through a gradual elaboration of logic. The multiple sources of error that are apparent here and which have caused confusion and conflict amongst philosophers can only be eliminated through increasingly better knowledge of logic. The main purpose of this course of lectures is to make clear in detail how important logic is for the development of philosophy in general.

If a teacher of philosophy should want to construct a philosophical system in accordance with the rules of a correct and fully developed logic, and then expound its contents to his students, it would be of very little use to them. They would only be adopting its results dogmatically and without further thought. What matters is the art of reasoning for ourselves and the art of discovering what is true through our own reasoning. In order to achieve this we must learn to avoid all the errors that lurk here; and for that we must study the sources of these errors and through practice gain multidimensional and comprehensive training in how to avoid those errors. Only in this way will we learn to recognise and see through all those mistakes in reasoning which lead us into error in philosophical matters. That is why logical practice for the sake of becoming acquainted with the relevant errors must come first, before it makes sense to begin with the positive construction of a philosophical system.

Lecture II

Practice in the art of avoiding the errors which lurk in philosophy is, I think, the purpose of a philosophical propedeutic, when conducted correctly.

In these lectures I wish to exhibit the importance of logic only from this critical perspective. These are lectures in applied logic, in the philosophical use of logic, or in the art of applying logic to the problems of philosophy. Therefore, what we are doing here is the identification and critique of dialectical illusion in philosophy. The importance of this task, particularly for philosophy, is clear from what I said earlier. Yet it is not self-evident that the identification and critique of dialectical errors is significant for any given subject. A logical critique of this type is unnecessary for fields of research that derive their results from the sources of intuition, or at least it does not have anything like the importance and significance that it does in philosophy. For in those fields we have intuition, which yields sufficient clarity about its objects. The sources of dialectical error are less dangerous there. Thus, the fact that philosophy does not have the same source of knowledge as those fields of research that do actually rely on intuition, is what makes logical critique so important in philosophy.

I have not wanted to speak of fallacies in general, but rather of *typical* fallacies in philosophy. At first sight, it might appear an impossibly large task to set forth what are the sources of all dialectic illusion in philosophy. For even if there is only *one* truth, the possibility of error seems limitless. It is therefore highly significant for us that this is *not* the case. It is not the case for we can show specific types of errors of reasoning that are typically found in philosophy, i.e. that occur again and again, so that the multiple errors found in philosophy can be reduced to a definite and indeed quite small class of errors. The same sources of error, few in number, are to be found in a multitude of applications. Hence, if we investigate the sources of illusion in philosophy we arrive at a *theory of dialectical error*, as I would like to term this exercise—a theory that allows for an overview of possible sources of error in philosophy and that provides us with a complete systematisation of the dialectical errors that originate from these sources. That is what I want to achieve in these lectures.

To elucidate the various typical fallacies I will present examples, and, wherever possible, examples taken from recent and even contemporary philosophy, since this is most familiar and important to us. To avoid any kind of misunderstanding I would like to state that I am not driven by an interest in a general critique of the philosophers from whom I will derive such examples. I beg you not to misunderstand my particular criticisms by generalising them beyond the specific points my examples touch on. By using real examples for our investigations we will disarm any worries that the theory targets straw men and deals in futile and unproductive quibbles. Only someone who has become privy to the dangers that lurk in the field of dialectics, as shown in my examples, and who, on the basis of such practice, has familiarised himself with the aforementioned typical sources of error, will be sufficiently aware of the dangers of dialectical illusion and equipped to undertake the search for a positive solution to the real problems of philosophy.

In my last lecture I spoke of the relativism which has of late widely shaken belief in philosophical truth. I then said that I did not want to embark on a refutation of relativism at this point. If I really succeed in fulfilling what I have stated as the aim of these lectures, then we will be indirectly in possession of a vastly more effective argument against relativism. If we succeed in getting all those disputes among philosophers that have led to scepticism reduced to a limited number of demonstrable errors of reasoning, then those disputes will cease to exist and with them the main fact which gives rise to relativism. Furthermore, the same result will highlight the benefit of such *logical* investigations, a benefit which is so little appreciated today. It will be shown that mere systematic elaboration of logic, once it is taken as far as it can go, is enough to put an end to the familiar disputes in philosophy and to secure the success that is denied to those who try to get there in their own sophisticate way, by skipping the strains of reasoning.

But at this point these are merely promises. We will allow the *outcome* to decide. I will for the moment only guard myself against a misunderstanding that could easily result from what I just said. There is, to be sure, a danger in *over* estimating mere logic and therefore reasoning. I have not in any way attempted to deny this. However, what I would like to draw your attention to is that a consistent development of the methods of our reasoning will guard us against that danger. It is after all only by misusing the methods of our reasoning that they can become dangerous for our pursuit of truth. There is, or at least there was in the past, widespread superstition surrounding the omnipotence of logic. That superstition does not take into account the fact that reasoning can indeed be correct in its logical form whilst its claim to be true is unjustified. An error can easily come about by expecting too much from logic, that is, when the relationship of the logical form to the content of the reasoning is misinterpreted and the mere logical form is considered sufficient to certify the truth of the outcome of an argument. This false expectation has itself derived from an error in reasoning, one that logic can specifically demonstrate, and through this demonstration we can guard ourselves against misapplying the methods of our reasoning. In fact, we get our first glimpse of the utility of logical critique here, and in my next lecture I will begin with an example taken from the misuse of logic to point out a typical philosophical fallacy.

Reference

Birken-Bertsch, Hanno. 2006. *Subreption und Dialektik bei Kant: Der Begriff des Fehlers der Erschleichung in der Philosophie des 18. Jahrhunderts* [*Subreption and Dialectics: The Concept of the Substitution Fallacy in Eighteenth-Century Philosophy*]. Stuttgart-Bad Cannstatt: Frommann-Holzboog.

Lecture III

Abstract Although logical reasoning is necessary for finding the truth among opposing philosophies, it is possible to overestimate its importance. This happens when consistency in a philosophical system is wrongly considered to be its main or even only criterion. A great philosopher will always be prepared to sacrifice a principle which leads to a falsehood and thus to allow his system to become inconsistent. This is illustrated with examples drawn from ethics (Bentham and Mill) and the philosophy of science (Poincaré and Le Roy).

I concluded my last lecture by forestalling a misconception that results from *over*estimating logical reasoning. As I have already said, there is little danger that this high regard for logical reasoning will gain the upper hand. Nonetheless, a concern might arise that *I* am the one to overestimate it. I have, after all, heaped great praise on the art of logical reasoning, made grand promises concerning the productiveness of this art, and placed them at the forefront of this course of lectures. For this reason it seemed best—so as not to discourage anyone from partaking in it—if I were myself to begin this investigation by analysing the error produced by such an overestimation of logic. Even if this error is less common today, it can recur any time as fashion changes. For changes in fashion are always to be expected in this field, as I have explained. When *that* happens, unprepared minds will be all the more confused and led astray.

I have already argued that the best way of guarding against this danger lies in an even deeper development of logical reasoning. It is, after all, only an error in applying logic that leads to a misinterpretation of its power and scope. Incidentally, even today this misunderstanding is also to be found in the specific shape of one popular opinion about what is valuable in a philosophical doctrine, namely that many people easily judge philosophical doctrines by consistency of structure, by the inner coherence of the system which is their outward expression. We often hear foolish praise being heaped on a philosopher who asserts the greatest absurdities yet remains faithful to his own principles and develops them consistently. This consistency, this coherence in the system, can all too easily seduce the critic into a false assessment of its content. In all this he forgets that consistency is only an empty form that in no way ensures the validity of the content.

For an unbiased judgment it is immediately clear that where there is a mistake in the fundamental premises of a philosophical system the consistency in the structure of the system can only serve to transmit this mistake to the entire system and so render it useless and without value in any of its conclusions. In contrast, an inconsistency can preserve the validity of the results of a philosopher who proceeds from false premises, for that inconsistency can help eliminate the original error.

There is another way of looking at the overestimation of consistency. It is easy to think that, if a philosopher develops his system from first principles, and he goes about this with exclusive attention to consistency, then this shows that the way he actually arrived at his conclusions was by first establishing his principles and then investigating what follows from them. This view entails that philosophers really do examine the truth of an assertion based on the consistency with which it follows from first principles.

In fact things are very different. The view just described is altogether a layperson's view. Not one important philosopher has approached his work in this manner. Consistency and coherence of the system, albeit overwhelming to laypeople, are always the last thing and never the starting-point for the creator of the system, at least if we're talking about an important philosopher. Things are quite the other way around—an important thinker will test the value of his principle by seeing whether its consequences are correct, and will prefer to be inconsistent than to relinquish a truth identified beforehand. There is no logical violence a philosopher would not be willing to inflict on his own system if the consistent elaboration of that system would lead him to a false result. A philosopher of note would always put his interest in truth above his interest in consistency.

In my last lecture I expressed the reason for this *prima facie* paradoxical yet indubitable fact—that the general principles in philosophy become clear to us much later than their particular applications. It is the feeling for truth that immediately connects with these particular cases; and it is from such a feeling that we (by means of abstraction) ascend to the underlying presuppositions. Only through such an analysis of the presuppositions of our factual judgments do we become aware of those general principles that are later placed at the forefront of the system. This is the reason for the layperson's impression that those principles were actually the starting-point of the philosophical system-builder, in fact what led him to building his system in the first place. It is only ever the insignificant minds in the history of philosophy, the epigones, so to speak, the followers of the great masters, who allow the consistency of the system to be the guiding star of their philosophical endeavours. This often results in the fact that by hammering away at the consistency of the system, they certainly make it increasingly consistent yet also increasingly barren and often increasingly absurd.

I would like to give a few examples of this to illustrate what I have been saying. My first example will be the ethical theory that harks back to the English school of the so-called Utilitarians. The founder of this school of thought was Jeremy Bentham, who lived at the beginning of the nineteenth century. He became famous not only in the history of philosophy, but also in the history of the social sciences. This philosopher established the principle that THE DESIRABLE is *PLEASURE*, so that the ethical value of an act depends on the amount and intensity of the pleasure that results from it.[1] This is the principle of what is called a hedonistic ethic. The originator of this system was of the opinion that the principle that GOOD = PLEASURE is not only a universally valid proposition, but also a self-evident and immediately admissible one, a proposition which does not have to be proved, so much so that he directly proceeded to derive consequences from this principle.

If someone really did want to make this the guiding principle of his behaviour, he would soon recognise that it conflicts with the endeavours of others who also act in this way, who also equate THE GOOD with PLEASURE. The desire for the greatest possible pleasure leads, as soon as it becomes general, to conflict. The increase in the pleasure that one person achieves is at the cost of the pleasure of others in society. The question is: how are we to apply that ethical principle in such cases? According to the principle we would be commanded to increase our own pleasure at the cost of others in all cases of conflict. But that cannot be an ethical command, for it is considered unethical to pursue our own pleasure at the cost of other members of society. Bentham certainly flinches from rejecting this generally accepted ethical proposition, according to which it is unethical to pursue our own pleasure at the cost of others. So what else can he do if he wishes to cling to his principle? There is nothing left for him except to introduce a new assumption, so as to bring consistency into the system, i.e. in order to bring the principle into agreement with the generally accepted proposition that it is unethical to increase our own pleasure at the cost of others. He must therefore introduce an arbitrary assumption, arbitrary for the simple reason that it does not proceed from his principle. He is forced to introduce an a priori axiom, by which the conflict in question cannot actually arise, whereby it is impossible that someone who only follows his own well-considered interest injures the interests of others. In brief, he assumes that it is worthwhile in the long run for the individual to value the interests of his fellow man as he would value his own, and so to abandon his pleasure wherever this would result in a decrease of pleasure in the community.

Thus, he is led to establish, as an ethical proposition, that the greatest happiness of the greatest number is the ethical ideal, a proposition which incidentally, as I said, not only rests on an arbitrary assumption, but has other logical weaknesses as well. For it unites two irreconcilable consequences. The greatest happiness for the greatest number is too big a requirement, for nothing prevents that under certain circumstances the only way to increase happiness might be to decrease the number of individuals to whom it is distributed; and nothing prevents either that an increase in that number might not in fact bring about a decrease in the happiness to be distributed to the individuals. This principle requires two *maxima* that are

---

[1] See Bentham (1789, Chap. 1). As explained in the Note on the Translation, SMALL CAPITALS indicate that either Nelson or the philosopher being discussed by Nelson (Bentham, in this case) is talking about *concepts*. These small capitals are not in the German original. Their purpose here is to highlight the frequency and centrality that focusing on concepts has in Nelson's text.

independent of each other.² But let us leave this objection aside. What I am concerned with in this case is the fact that this philosopher, in order to bring his dogmatical principle into agreement with a particular truth that he does not dare to doubt, introduces an arbitrary assumption that does not follow from his principle and indeed is clearly false.

Bentham's mistake was noted by his follower, John Stuart Mill, who saw that he had to correct the doctrine of his master.³ His correction consisted in the following thought: if it is unethical unconditionally to pursue the increase in one's own pleasure, then this suggests that there is another form of valuing pleasure than simply measuring its strength. There are different *kinds* of pleasure. There is a *higher*, a *true* happiness, as expressed by the fact that it is to be considered more valuable even if the pleasure that it consists in is smaller. It is, says Mill, clearly

> better to be a human being dissatisfied than a pig satisfied; better to be Socrates dissatisfied than a fool satisfied.

Hence it is not enough to take into consideration merely the magnitude of the pleasure, it is also important whether it is a higher or a lower kind of pleasure.

This sounded so simple that Mill was wholly reassured with the change and believed that it had saved his teacher's original principle. As a matter of fact, an impartial analysis will show that he had thereby completely destroyed this principle. He did not notice that the consistency of the system was thereby entirely shattered, for the height of the pleasure that is in question here presupposes a criterion utterly outside of all pleasure. How could we otherwise assess the different kinds of pleasure according to whether they are higher or lower? A pleasure according to this new conception is good when it is directed towards something good; the criterion of the good therefore lies outside of pleasure and is independent of all pleasure. In fact, this criterion restricts the desire for pleasure.

Let us consider another example, this time taken from the field of theoretical philosophy. The great mathematician Poincaré (1902) developed a new, ingenious theory of the origin of geometric and physical principles. According to his view, the origin of these principles is to be found in a place entirely different from the one envisaged by those who had previously investigated the question. He believed that it is not to be found in any kind of knowledge but rather in mere stipulations, in hidden conventions. He did not just assert his view but also argued for it. His argument consists in showing that those who had up till then believed that the origin of these principles lies in logic are as wrong as those who had supposed that they have an experimental origin. The axioms of geometry, for example, cannot be established by pure logic. From the standpoint of pure logic these axioms are just as good as any other assumptions that contradict them. And they have no experimental origin, i.e. there is no experiment that can establish their truth. Experiments have to

---

²This is a mathematical objection. In calculus it is impossible to find a unique solution satisfying two (or more) maxima at the same time. See Edgeworth (1881, Appendix VI).

³See Mill (1861, Chap. II).

Lecture III

be interpreted in order to draw conclusions of such scope, and this interpretation already presupposes the axioms. It is therefore impossible to want to establish these axioms through experimentation. Observed or experimental facts are as much in agreement with one axiom as with its negation. The origin of these principles lies neither in logic nor in experience—and the same is true of the general principles of physics.

In this way, Poincaré argues for a doctrine according to which the origin of these principles must lie in stipulations or conventions, which could just as well have been different, and a choice between these is a matter of convenience not truth.[4]

Now, this theory plays an important role in the dispute over the value of science. What will become of the value of science if its general principles are merely conventions?

It was a disciple of Poincaré himself, the mathematician Édouard Le Roy, as important a mathematician as he was a devout Catholic, who drew the conclusion, on the basis of his master's teachings, that the laws of nature are not true, and that therefore the old dispute between natural science and dogmatic Christian theology had been finally decided in favour of the latter.[5] Natural science can no longer lay claim to imparting truth through its so-called natural laws. Anyone who seeks to be taught the truth must appeal to another source. Truth only flows within the traditions of the church. Poincaré, somewhat dismayed by the conclusion his disciple had drawn from his own doctrine, attempted to protect his science against this attack from dogmatic theology. He was in a difficult position, since consistency was clearly on the side of his follower Le Roy.

How then did Poincaré attempt to avoid the said conclusion? He said: the truth, for which Galileo had suffered, the truth that the earth moves, is as firm as ever.[6] What does it mean to say that such a proposition is true? It means nothing other than that it is a convenient assumption, and it *is* convenient to assume that the earth moves. The assumption turns out to produce an astonishing number of results, and this shows it is convenient.

This appears at first sight to be highly reassuring, and Poincaré certainly felt reassured by this solution to the question. But the unbiased critic will soon

---

[4]Poincaré's conventionalism is further discussed in Chapters "Lecture X" and "Lecture XI". For a modern exposition of this doctrine, see Ben-Menahem (2006).

[5]See Le Roy (1899, 1900).

[6]Poincaré's French has, *La terre tourne*, which is quite accurately rendered by Nelson's German, *Die Erde dreht sich*. As any schoolchild has been taught, the earth has two movements, one of rotation and one of translation. The French and German sentences mentioned are neutral as to which of these two movements is meant. The English rendering that best keeps this ambiguity is, 'The earth moves' (as in Galileo's famous *eppur si muove*). Incidentally, the discussion in Poincaré's text develops his point first with respect to the first kind of movement ('The earth rotates') and then with respect to the second one ('The earth orbits').

recognise the weakness in this argument. We only need once to ask ourselves: which truth is it that, according to this assertion, Galileo suffered for? I shall try to write it down. The proposition, states Poincaré, that the earth moves has no other meaning than that it is convenient to make the following assumption:

(1) It is convenient to assume that the earth moves.

But *what* assumption is convenient according to (1)? What does it mean then that the earth moves? It is written right there in (1). The assumption means that it is convenient to assume that the earth moves. If we therefore express ourselves more explicitly in rendering our thought, we will have to include this explanation and state the proposition (2):

(2) It is convenient to assume that it is convenient to assume that the earth moves.

And if we now inquire further what is the assumption that it is convenient to make, we are given the same explanation all over again—that it is convenient to assume the convenience of this last assumption. We thus arrive at an infinite series of explanations when asking for the meaning of our proposition, each new explanation deriving its meaning only from the next one. And because the series is infinite there is no way we can ever grasp the meaning of the original assertion. If the proposition that the earth moves means that it is convenient to assume that the earth moves, then it is utterly certain that the proposition does not have any meaning at all.

There is the same lesson in this example as in the first one: the only function of a philosophical principle lies in justifying afterwards a truth that the philosopher has assumed in advance. If the principle cannot fulfill this function, then a philosopher worth his salt does not pursue consistency for its own sake by continuing to draw conclusions from that principle, since he would have to declare false what is obviously true. Instead, he modifies his principle in order to save a standing truth, even at the cost of some inconsistency in his system. Le Roy was more consistent, but Poincaré was a better philosopher.

We can see from this example that the consistency and truth of a system are two different things. If consistency was the only important aspect, then it would be just as easy to make the system consistent by keeping the principle and giving up the feeling for the truth (the latter vouching for a particular truth), as the other way around, by keeping the feeling for the truth and giving up the principle. Even this simple consideration proves that logical consistency is not a sufficient criterion of truth. If we were to judge only in accordance with consistency, then one system would be just as good as another, either to keep the principle and sacrifice the particular truth or to keep the latter and sacrifice the former. Both solutions are equally consistent. The demand for consistency remains ambiguous because both systems satisfy it equally well. Nevertheless, the two systems contradict each other. Therefore, if the criterion of truth is consistency, a contradiction would be true, which is against the logical principle that underlies the demand for consistency itself, i.e. the principle of the exclusion of contradiction. Choosing the true system means evaluating the principles themselves—as opposed to simply showing that the system is consistent. If it was a matter of evaluating these principles *logically*, then we would have to reduce them to higher principles, for no proof is possible without presuppositions. Yet the same thing is true of these presuppositions. They can only be logically evaluated by means of even higher ones from which they are derived, and so on ad infinitum. Hence, if we actually wanted to evaluate the truth logically and so to provide a proof of all truth, then we would have no starting-point for proving anything. Hence, no proof and therefore no truth would be possible, and certainly not even the truth that only what can be proved is true.

References

Ben-Menahem, Yemima. 2006. *Conventionalism*. New York: Cambridge University Press.
Bentham, Jeremy. 1789. *An Introduction to the principles of morals and legislation*. London: Payne.
Edgeworth, Francis Y. 1881. *Mathematical psychics: An essay on the application of mathematics to the moral sciences*. London: Kegan Paul.
Le Roy, Édouard. 1899, 1900. Science et philosophie [Science and Philosophy]. *Revue de métaphysique et de morale, VII*, 375–425, 503–562, 708–731; *VIII*, 37–72.
Mill, John Stuart. 1861. Utilitarianism. *Fraser's Magazine* 64: 391–406, 525–534, 658–673. [Reprinted many times].
Poincaré, Henri. 1902. *La science et l'hypothèse*. Paris: Flammarion. [English Trans. *The Foundations of Science: Science and Hypothesis, The Value of Science, Science and Method*, New York, The Science Press, 1929, pp. 9–197].

Lecture IV

Abstract The excess of confidence in logic culminates in *logicism*, a position common to medieval Scholasticism and modern rationalism. This mistake can best be illustrated by the idea, especially developed by Leibniz, that the lack of contradiction in a concept is a warrant that the corresponding object exists. Certain inconsistencies in Leibniz's system were corrected by Wolff, whose excesses finally allowed Kant to uncover the logicist fallacy.

In my last lecture I addressed the error that consists in *over*estimating the value of logical reasoning as expressed by letting consistency be the sole criterion for judging a system of thought. The strict bounds of logical reasoning, dictated by its mediating character, are thereby neglected. To say that pure logical thinking 'is mediating' means that the *content* of truth cannot be generated by the pure *form* of logical inference, albeit logical inference can indeed help establish what follows from this content as taken from somewhere else. Logical reasoning is in and of itself *empty* in the sense that it cannot create its own epistemic content.

The insight that rational knowledge depends on something else is not so old. The error I have described did not only entirely pervade the medieval philosophy of the *Schoolmen*, but also the philosophical school which under the name of *rationalism* dominated the modern era up to the end of the eighteenth century. Only towards the end of the eighteenth century was the spell of this conception broken, with the arrival of Immanuel Kant. But even Kant's arrival was not able to prevent the recurrence of that error, which even today emerges here and there.[1]

What characterises scholastic philosophy is the conception that it is possible to answer scientific questions through logic. And this characterisation of the Schoolmen's fundamental view allows us to say that rationalism is the heir of

---

[1] The reader will remark, from this and many other passages of this work, that Nelson thought very highly indeed of the contributions made by Kant towards the identification and prevention of typical fallacies in philosophy. This might lead the reader to expect that Kant was, in Nelson's view, somehow immune to these fallacies. To show this is not at all the case a brief sample of Kantian fallacies, as identified by Nelson himself, is given in the Appendix.

scholasticism and does not differ from it in any fundamental way. There is nonetheless a difference, and one rich in consequences, but it lies elsewhere. Rationalism broke off from the tyranny of church dogma. It freed scholasticism from being a maid to theology. Yet this difference shows the error in a more glaring light. For indeed logic was now not only charged with the task of providing an independent defence of church dogma, but also the additional charge of obtaining by pure logical inference a transcendent truth high above experience.

Typical statements made by well-known thinkers can easily provide evidence of this. For example, Blaise Pascal stated that every truth must be capable of being proved and that if this had not been achieved by man up to now, the reason could only be attributed to the limitations of human understanding.[2] The consequences of this error are well known—in the guise assumed in all teachings of this philosophy, i.e. as theorems proved in the mathematical manner. Universal principles (called 'axioms' and 'definitions') are placed at the top in order to derive the entire system from them. Spinoza's ethics has become the best known example of a project of this sort, which even today enjoys some popularity. Spinoza endeavours to derive the ethical truths *more geometrico*, as he says. The word 'rationalism', which is used to describe this tendency, is not in fact appropriate. That word means the opposite of experience or empiricism. It does not express clearly enough the fundamental view of this way of doing philosophy. For what is needed is to say positively what is the source of knowledge which alone the philosopher intends to draw from. This source of knowledge is not sufficiently determined by saying it is *not* experience, but only by saying positively that the philosopher intends to draw only from logic, from pure reasoning. Hence, it would be better to call this school *logicism*.[3] This term makes clear that it is in fact just a new scholasticism.

Let us look at the error of logicism more closely. We will then be all the more protected against reverting back to it. First we shall look at the basic problem that must unavoidably be at the centre for scholastic and rationalist philosophers. This problem arises automatically if these philosophers aim at gaining knowledge by pure reasoning. They *will* view universal concepts as the foundations for the knowledge of everything that exists; in other words, they *will* deem it possible to grasp objects, existing things, on the sole basis of universal concepts. The problem is therefore how to go from universal concepts to objects, to existing individuals.

---

[2] See Pascal (ca. 1657).

[3] At the time of these lectures it seems that the term 'logicism' had not yet been taken to refer to the doctrine, most closely associated with Frege and Russell, according to which the whole of mathematics could be reduced to logic; see Grattan-Guinness (2000, 3–4, 479 and 501), whose account of the complicated history of the doctrine itself is authoritative. In that sense, Nelson's usage is probably pioneering, although he clearly meant it much more broadly and polemically (see already Nelson 1908, 1917, where he applies it to epistemology and ethics, respectively).

The main feature of a CONCEPT is its universal character, whereas that of an EXISTING OBJECT is its individuality.[4] Philosophers therefore looked for what they called a universal *principium individuationis*, a principle that was to be universal in itself yet capable of picking out the individual. How on earth could *that* be possible? If philosophers take the lack of contradiction as a sufficient criterion of truth, or more precisely if they take the mere lack of contradiction of a concept as a sign that something actually exists, then they have to arrive at the principle that *predicates do not oppose each other*.[5]

This principle plays a particularly important role among the followers of Leibniz. For what is a contradiction? It is nothing other than an attribute that both does and does not belong to an object. There is then a contradiction only between the position and the negation of an attribute, but never in the reciprocal relationship of positive attributes. Positive attributes can never contradict each other. This is how those philosophers arrive at the principle, 'Predicates do not oppose each other'. They arrive further at the notion of a UNIVERSAL SET, i.e. the set of all positive attributes.[6] All positive attributes of things must be compatible with each other according to that criterion, and a contradiction can never arise between any two of those predicates. The question is then: how is it possible that despite this, as experience shows, certain predicates nevertheless exclude each other in existing individuals, e.g. that a winged horse is impossible, despite the lack of contradiction in the concept?

There is a problem here if we proceed from the purely logical criterion of truth, which in itself only excludes contradiction. It therefore becomes a task for this philosophy to explain the possibility of existing individuals on the basis of the

---

[4]At this point a theme is announced that will recur in these lectures. Fries and Nelson, following Kant's strictures, and in opposition both to German idealism and to German (nonmathematical) 'logicians', distinguished terminologically between a concept (*Begriff*) and an object (*Gegenstand*), very much in the way Frege (1892) later did. It is a shame that Nelson died before Frege's logical views and the much more perspicuous notation they allowed became generally accepted in Germany, to a large extent thanks to the first modern logical treatise (Hilbert and Ackerman 1928). If readers keep this in mind, they will understand certain later passages, especially in Chapters "Lecture V", "Lecture VI", "Lecture VII", "Lecture XII", "Lecture XV", "Lecture IX" and "Lecture XXII". See Footnote 6 in Chapter "Lecture XXII".

[5]See Kant's essay on 'the only possible argument for the existence of God' (1763, Sect. 1, Third Reflection, §6; English version in Walford 1992, 130). The word 'predicate' translates the German word *Realität*, which cannot be translated by 'reality', for in the scholastic tradition, to which the Leibnizo-Kantian school terminologically belongs, *Realität* does not refer either to actual existence or to actually existing things, but rather to the positive features or properties of an object that itself might or might not exist (see Wolff 1736, §243). This is why the modern logical term 'predicate' is particularly apt in this context.

[6]The German phrase is *All der Realitäten*, and here again the word *Realität* is a higher-order term referring to any positive predicate, to a quality or a property that an object may have or lack. Although the phrase *das All der Realitäten* (or rather *das All der Realität*) stems from Kant, the concept is thoroughly Leibnizian. See Kant, *Critique of Pure Reason*, A575–576 B603–604, A628, B656. The expression 'Universal Set' seems to render the meaning intended in a way that is clearer for contemporary readers.

presupposed Universal Set. The Universal Set must somehow be restricted, for only some particular predicates are assigned to existing individuals to the exclusion of many other predicates. The reason for this exclusion is only to be found in a restriction of the Universal Set; to say it with Spinoza: *Omnis determinatio est negatio*.[7] The predicates that belong to the existing individual are those that remain after some of them (as united in the Universal Set) are negated of this particular thing. Or to use the words of those philosophers, in essence all things are unified in God, for God as the most perfect being is nothing other than the Universal Set.[8] God in His perfection cannot lack any predicate. However, with regard to *existence*, things lack quite a few predicates as they are unified in the Universal Set—in God. The set of all predicates is impoverished as it passes from essence (in God) to existence. Not all predicates in the realm of existence are *compossible*, again to use the way of talking of the Leibnizians. Now what is the criterion of compossibility?

It is easy to recognise the contradiction intrinsic to the task these philosophers set themselves. For if the law of non-contradiction is to serve as the sole criterion of truth, we will never arrive at the necessary restriction of the Universal Set that leads to existing individuals. Consistency must not be satisfied only by existing individuals but also by the Universal Set itself. The law of non-contradiction does not allow us to restrict the set of all predicates and so can never give us a *principium individuationis*. The conception of such a PRINCIPIUM INDIVIDUATIONIS contradicts itself, for the basis of *individualisation* cannot for logical reasons reside in a *universal principle*.

Spinoza says somewhere that a false assertion states something about its object that is not contained in our concept of it.[9] So an assertion about an object is true because of the mere *concept* of this object, and it is false because it contradicts that concept. Leibniz is also under the spell of this fundamental scholastic conception, which he developed more consistently. He went so far as to offer a proof, in a paper consisting of a series of sixty theorems, that only the Count Palatine of Neuburg was suited for election as the Polish King.[10]

However, Leibniz draws an interesting distinction in this context. He classifies all truths into *necessary* or (his expression) *eternal truths* and *contingent* or *factual truths*. The principle of the first class, the necessary truths, is just the law of non-contradiction. By contrast, the principle of factual truths lies for him in the 'principle of sufficient reason'. Factual truths are subject to hypothetical necessity, and one knows that an assertion is hypothetically necessary through the principle of reason. This distinction seems to draw a sharp limit to pure logic, factual truths

---

[7]In this form the proposition actually stems from Hegel, who was misquoting from a letter written by Spinoza to his friend Jarig Jelles (see Morgan 2002: 892) and probably being guilty of overinterpretation. A good discussion on the relevant issues is Melamed (2012).

[8]See e.g. Baumgarten (1779, §803).

[9]He says that in his incomplete and posthumous *Treatise on the Improvement of the Understanding*, §72. See Morgan (2002: 20).

[10]Although the story is decidedly funny, there may have been method in the madness. See Griard (2008).

Lecture IV

being at least excluded from its competence. But it only seems so. For Leibniz's explanation for his division of truths into two classes is that our ideas are not clear enough—only this justifies separating factual truths from necessary truths. The universal criterion of truth, according to Leibniz, is to be found in the proposition, *Praedicatum inest subiecto*, 'The predicate of the proposition is contained in the subject.'[11] So to prove that a proposition is true what we have to do is to explore the idea of the subject, to analyse that idea until we discover the predicate as one of its elements. Yet our ideas are in part so confused that we do not always manage to make this discovery. That is all there is to factual truths. Factual truths are true of objects about which we have ideas so confused that we do not manage to discover that the predicate of the statement we make is already in the idea of the subject. The lack of clarity of our ideas sets us limits, but the said distinction does not exist for an unconstrained understanding such as God's mind—for Him necessary truths and factual truths do not come apart, and His ideas are so clear and distinct that the predicate is always part of the idea of the subject. It must therefore be possible for God's perfect mind to derive all factual truths from the concept of the object by pure logic.

Christian Wolff, a follower of Leibniz, went even further. What Leibniz had ascribed only to God's mind Wolff tried to extend to the human mind. He tried to reduce the principle of sufficient reason (according to Leibniz the principle of factual truths) to the law of non-contradiction, the principle of necessary truths, thereby erasing the boundary between the two classes of truth. This proposal actually made Leibniz's conception more consistent. For if, as Leibniz taught, the principle of sufficient reason is the principle of factual truths, then this principle is not itself a factual truth, but instead a necessary one, as its universality already suggests. Therefore it should follow from the law of non-contradiction. Wolff tried to prove that this is so. He argues as follows: assume the principle of sufficient reason is not valid. Then we would have to admit that a thing's coming into being is possible without reason. That, however, would mean that something can come from nothing, which is evidently impossible. Hence, the assumption is false and the principle of sufficient reason proved.[12]

We can today see the error behind this proof quite easily, and the fact that a pseudo-proof like this could deceive a thinker like Wolff only shows how utterly the logicist prejudice dominated at that time. Again we have an example of the fact of which I spoke in my last lecture—an example of how a disciple of a great philosopher can forfeit all concession to truths external to his master's system in the interest of the consistency of that system. The disciple thus develops a truly unified system, thereby making its starting-point all the more obviously incorrect to the impartial observer. But this is also the reason for the significance of the kind of

---

[11] This principle is mentioned by Leibniz on different occasions and in various ways, most famously in his *Discourse of Metaphysics* and his correspondence with Antoine Arnauld. See e.g. Loemker (1969: 307, 310, 337).

[12] See Wolff (1736, §70).

achievement that we thank Wolff for in the history of philosophy. For success in uncovering and eliminating an error of this kind can only be attained when it has been consistently developed, so that its root lies clearly in front of us. In this sense, it can be said that Wolff was the first philosopher to have carried the logicism of the Schoolmen to its ultimate consequences. The error in this way of doing philosophy is finally revealed. The aforementioned proof is an example of this. The proof is a trick. The principle that is to be proved is a tacit premise of the proof. For the principle that something cannot come from nothing is merely a different formulation of the principle of sufficient reason that is to be proved. We call this type of fallacy a *petitio principii* or 'begging the question' (i.e. using as a premise the statement that is to be proved).

As I said earlier, we are grateful to Kant for finally overcoming the logicist fallacy. But we still find this error in Kant's first period of activity. It was with great difficulty that Kant was able to free himself from the spell of this way of doing philosophy. He was himself a disciple of Wolff's. In his early work he asserts that the law of non-contradiction is the criterion of all truth, which he formulates as 'no predicate belongs to a subject that contradicts it' (a formulation that actually only differs from Leibniz's *praedicatum inest subiecto* in being a negative statement).[13] According to this principle, a statement is recognised as true in that the negation of the predicate contradicts the subject. But clearly this proposition does not in any way offer a criterion of truth. It can at best be used as a mere nominal definition of truth.[14] For indeed a statement is true if what it predicates of an object actually belongs to the object. If the predicate did not actually belong to the object, we would just be making it up. But this does not give us a criterion at all. Such a criterion would allow us to *know whether* a predicate actually belongs to an object or not. Now, we can certainly know that a predicate belongs to an object if it belongs to the attributes that are contained in the *concept* of the object; and that is precisely what Kant had in mind when he formulated his law of non-contradiction. If he had said that no predicate belongs to a subject if it contradicts the subject's *concept*, then he would have formulated the law of non-contradiction correctly. The law of non-contradiction says that if one predicate negates another predicate that is contained in the concept of a given object, then the former predicate cannot be said to belong to that object. Yet the inverse is not true. Not every attribute that belongs

---

[13] See Kant's *Inquiry Concerning the Distinctness of the Principles of Natural Theology and Morality* of 1763, Third Reflection, §3 (English version in Walford 1992, 267–268).

[14] Every definition has the same linguistic form: it associates a chain of words (the *definiens*) with a given term (the *definiendum*). The association might be a pure (thoroughly arbitrary) stipulation, as when I say that, from now on, the invented word 'schmin' will mean 'a chimney of pyramidal shape'; or it might have some empirical value, as when a lexicographer explains the meaning that a given word (say, 'unicorn') has for a given cultural community. All such definitions are purely *nominal*, in that the definitions do not imply the *existence* of an object in the real world that the defined term refers to or of which the defined term might be predicated. The opposite of a nominal definition is a *real* definition, which either gives directions for the construction of a mathematical object or describes the operations by which we can identify or detect a thing or process. See Footnote 11 in Chapter "Lecture XI".

to the subject is *ipso facto* already contained in the *concept* of the subject, and not every negation of an attribute that belongs to the subject therefore already contradicts the concept of the subject.

Let us take for example the concept of a CIRCLE as a CURVE ON A PLANE WITH A CONSTANT DISTANCE FROM A FIXED POINT. If we state that all radii of a circle are equal, then the negation of the predicate would contradict the concept of the subject as we have just defined it, and this definition is the reason why our statement is true. Yet if we say, for example, that the circle is a closed curve, then we attribute a predicate to it that is not already contained in its concept. The concept does not in the least contain all attributes that belong to the object subsumed under that concept, but instead it only unites those attributes that are necessary and sufficient for assigning the object to a particular class.[15] That is precisely why the concept is not a sufficient reason if we want to know the positive attributes of the object, but it only suffices to *exclude* certain attributes of an object, i.e. those that contradict the concept.

The Schoolmen's endeavour to find a universal *principium individuationis* failed just at this point. The existing individual is not contained in the universal and cannot be derived from it by restricting the Universal Set—by negation, as Spinoza wanted. Hence, there is no transition from the universal concept to the existing individual. An unlimited number of individual things can still be subsumed under any given concept, irrespective of how specifically we might define that concept. The concept in and of itself does not decide in any way whether things and how many things are subsumed under it. To decide this it is necessary to have a source of knowledge different from the concept. The concept in and of itself only determines a *class* of objects and it only determines the individual object *qualitatively*, as far as those attributes belong to it that are necessary and sufficient for assigning it to that class. In other words, the concept never determines an object *numerically*, i.e. not as *this one* object, different from *that one*, even though both are perhaps qualitatively equal and belong to the same class. The existence of an object is not a qualitative feature of an object and cannot in any real sense be ascribed to it as a universal attribute. This was attempted in the ontological proof of God's existence —the purest expression of what the Schoolmen were up to. In it they tried to prove God's existence by using only the law of non-contradiction and the definition of GOD as THE SUPREMELY REAL BEING, i.e. THE BEING WHO CANNOT LACK ANY ATTRIBUTE, including the attribute of EXISTENCE. For the negation of the attribute of EXISTENCE would contradict the concept of such a Being, and EXISTENCE is an attribute intrinsic to His concept.

---

[15]The phrase 'necessary and sufficient' refers to a biconditional. Nelson is talking about a purely logical constraint for definitions, viz that every definition has the form 'x is F if and only if x is G', where F is the concept to be defined and G is the concept that defines F. Given any concept F that we want to define, there is usually no unique way to define F. Mathematicians and scientists in general *choose* among an indefinite number of equivalent definitions the one most suited to the purpose at hand. Observe also that the whole purpose of a definition is to enable anybody to pick up the *object* denoted by the term defined. Definitions are for Nelson, as for most contemporary analytic philosophers, extensional.

References

Baumgarten, Alexander G. 1779. *Metaphysica*, 7th ed. Magdeburg/Halle: Hemmerde.
Frege, Gottlob. 1892. Über Begriff und Gegenstand. *Vierteljahresschrift für wissenschaftliche Philosophie* 16: 192–205. [English translation: On Concept and Object, *Mind* 60(238): 168–180, 1951].
Grattan-Guinness, Ivor. 2000. *The search for mathematical roots, 1870–1940: Logics, set theories and the foundations of mathematics from Cantor through Russell to Gödel*. Princeton: University Press.
Griard, Jérémie. 2008. The *Specimen demonstrationum politicarum pro eligendo Rege Polonorum*: From the concatenation of demonstrations to a decision appraisal procedure. In *Leibniz: What kind of rationalist?* (ed.) Marcelo Dascal, 371–382. Dordrecht: Springer.
Hilbert, David, and Wolfgang Ackermann. 1928. *Grundzüge der theoretischen Logik*. Berlin: Springer. [English translation of a subsequent edition: *Principles of Mathematical Logic*, New York, Chelsea, 1950].
Loemker, Leroy E., ed. and transl. 1969. *Leibniz: Philosophical papers and letters*. Second edition. Dordrecht: Kluwer.
Melamed, Yitzhak Y. 2012. 'Omnis determinatio est negatio': Determination, Negation, and Self-Negation in Spinoza, Kant, and Hegel. In *Spinoza and German Idealism*, (eds.) Eckart Förster, and Yitzhak Melamed, 175–196. Cambridge: University Press.
Morgan, Michael L. (ed.). 2002. *Spinoza: Complete works*. Indianapolis: Hackett.
Nelson, Leonard. 1908. Über das sogenannte Erkenntnisproblem [On the So-Called Problem of Knowledge]. *Abhandlungen der Fries'schen Schule (N.F.)* 2(4): 413–818. [Reprinted in Nelson (1971–1977), vol. II, pp. 59–393].
Nelson, Leonard. 1917. *Kritik der praktischen Vernunft [Critique of Practical Reason]*. Göttingen: Vandenhoeck & Ruprecht. [Reprinted in Nelson (1971–1977), vol. IV].
Nelson, Leonard. 1971–1977. *Gesammelte Schriften*, 9 vols. Edited by Paul Bernays, Willy Eichler, Arnold Gysin, Gustav Heckmann, Grete Henry-Hermann, Fritz von Hippel, Stephan Körner, Werner Kroebel, and Gerhard Weisser. Hamburg: Felix Meiner.
Pascal, Blaise. ca. 1657. De l'art de persuader. Manuscript published for the first time by Pierre Nicolas Desmolets in *Continuation des mémoires de littérature et d'histoire* 5(2): 271–296. Paris: Nyon. [English translation: The Art of Persuasion, in *Pensées and Other Writings*, pp. 193–204, Oxford, Clarendon Press, 1995].
Walford, David, ed. and transl. 1992. *Immanuel Kant: Theoretical Philosophy, 1755–1770*. New York: Cambridge University Press.
Wolff, Christian. 1736. *Philosophia prima, sive Ontologia, methodo scientifica pertractata [First Philosophy, or Ontology, Developed by the Scientific Method]*. New revised edition. Frankfurt and Leipzig: Officina Libraria Rengeriana.

Lecture V

Abstract The ontological 'proof' of God's existence is an argument which best shows what the logicist fallacy is about. Kant dissolved this 'proof' as well as many other logicist arguments by means of his distinction between analytic and synthetic judgments. His presentation of that distinction has its flaws, but the distinction as such can be defended against all objections.

In my last lecture I talked about a logicist prejudice and how it dominated the philosophy of the Schoolmen and its heir in the history of philosophy, the so-called rationalist school. These philosophers place their trust entirely in logic, in rational knowledge, for they believe that by a long enough series of conceptual distinctions the particularity of individual existing things can be derived from the universal without the aid of intuition. This prejudice finds perhaps its most definite expression in Leibniz's so-called *principium identitatis indiscernibilium* (principle of the identity of indiscernibles). This just means that a complete definition of the concept guarantees the identity of the object. Thus if we are dealing with several existing things, then they must have different attributes. For it is by these attributes that the objects are conceptually definable. So it must be enough to have defined an object by all its qualities to be certain of the identity of the existing object defined by this concept. It would indeed have to be like this if knowledge of objects was possible by pure reasoning. Existing things would then be different from each other if and only if one is F and the other not F relatively to whatever concept F we use, according to the logical principle that everything either falls or does not fall under a given concept.[1]

We discovered that the error in this view lay in the fact that its adherents had failed to recognise the nature of the concept. The concept is and remains a general attribution. In itself the concept only determines a class of objects, never an individual. Even if we think of the individual as fully determined by its qualities, it would still of course be possible that other individuals, *numerically* distinct, *several*

---

[1] The German way of indicating the relation between a concept C and an object O, of which C is predicated, is to say that O 'falls under' C. Although not idiomatic in ordinary English, the influence of Frege and his translators has made it so familiar to philosophical readers that it is retained throughout in this translation.

of them, fall under the same concept, i.e. there can be more than one qualitatively identical individual. For an existing object is not only qualitatively but also numerically determined, as for instance when we distinguish one object from another by their locations in space. However, this numerical determination is not part of the concept and requires intuition. As I argued in my last lecture, failure to recognise this is what explains the error in the ontological proof of God's existence, this masterpiece of Scholastic dialectics. The ontological proof proceeds by forming the concept of THE MOST PERFECT BEING, a being so perfect that it unites all predicates in itself, which it can do because (according to the Schoolmen's presupposition) predicates do not contradict each other. Predicates do not oppose each other if contradiction is the only criterion of opposition. Contradiction is only possible between an attribute and the negation of that attribute, never between two positive attributes. Now (so the Schoolmen's reasoning continues) the most perfect being would not unite all predicates in itself if it lacked the attribute of EXISTENCE. We cannot therefore deny that being's existence, without committing a logical contradiction.

The error in this proof is that existence cannot be part of the concept of an object the way all general attributes can be—as Kant, whom we have to thank for clarifying this matter, at one point wittily notes: one hundred existing dollars are not more than one hundred possible dollars.[2] The only difference is that the former do exist. It is the *same concept*, whether or not there is actually an object that falls under it. For whether an object exists that falls under the concept cannot be known on the strength of that concept alone.

Now, it is perfectly possible to construct a concept of EXISTENCE, and it is then entirely possible, as a matter of formal logic, to predicate of an object the attribute of EXISTENCE, so that the attribute of EXISTENCE becomes part of the content of another concept.[3] It would indeed follow, lest we contradict ourselves, that every object that falls under this concept also exists. But we have to be clear about what we are really saying here. What does it mean to say that every object falling under that concept also exists? It means that an object exists which has the attributes contained in the concept in question, one of which is the attribute of EXISTENCE. In order then to subsume any object under this concept, we must know in advance that the object exists, and we have to know this for other reasons than the mere knowledge of the concept in question. Once we subsume the object under the concept as a result of such knowledge, we can of course derive again its existence from the concept. But no new knowledge is thereby obtained. We only arrive at the trivial hypothetical proposition that if an object which in addition to this or that other attribute also has the attribute of EXISTENCE, then such an object exists. In order *categorically* to assert that something exists we need non-conceptual reasons.

---

[2]See *Critique of Pure Reason*, A599 B627.
[3]Nelson shows himself quite prescient here, for this was in fact done within *free logic*, one of the non-classical systems of logic (see e.g. Lambert 2002, Chap. 2).

Kant coined a technical expression for this result on the basis of his distinction between analytic and synthetic judgments, the sole distinction capable of clearing up the misunderstanding. Existential propositions are always synthetic judgments, i.e. what they assert cannot be derived from the mere concept of the object. This distinction between analytic and synthetic judgments is so momentous that it repays closer attention. You will remember the proposition that Leibniz formulated as the general criterion of truth: *praedicatum inest subjecto*—the predicate is contained in the subject. I have shown you that this proposition does not provide a criterion of truth, but only a mere definition of the concept of TRUTH. An assertion is true if and only if the predicate actually belongs to the object to which it is ascribed. Otherwise we would be dealing with a pure fabrication, we would be ascribing an attribute to an object to which it does not belong, thereby asserting something that is not true. However, we can never conclude from this proposition whether a certain attribute belongs to it or not. This error stems from mistaking the subject for the concept of the subject. But we do obtain a criterion of truth if we use the concept of the subject rather than the subject itself. If we ascribe an attribute to an object that is already contained in the concept of the subject, then we can be sure that our statement is true. But if we only use such predicates, then we can never expand our knowledge of the object beyond the set of properties included in its concept. Judgments that say something about a subject that is already contained in its concept are termed *analytic*, because they just analyse the concept of the subject. Judgments, on the other hand, that go beyond the set of predicates included in the concept of the subject we term *synthetic* because they join a new concept with the concept of the subject and thereby expand our knowledge of the object beyond the mere concept. For example, the proposition, 'All bodies are extended', is an analytic judgment because the attribute of EXTENSION is already contained in the concept of BODY. By contrast the proposition, 'All bodies gravitate', is a synthetic judgment because GRAVITY is a property that is not already contained in the concept of BODY. We must go beyond the mere concept of BODY and look elsewhere in order to discover that GRAVITY is a property that belongs to bodies.[4]

The connection here to the law of non-contradiction, which according to the presupposition of logicism is what all truth must be reduced to, is obvious. Using the law of non-contradiction we can establish the truth of all *analytic judgments*. We cannot deny an analytic judgment without contradicting the concept of the

---

[4] Kant's example of a synthetic judgment or proposition, *Alle Körper sind schwer*, is apparently easy to translate as, 'All bodies are heavy'. However, we must remember that Newton's physics was always in Kant's mind. And from this perspective a much deeper translation would be, 'All bodies gravitate'. This is an astonishing synthetic proposition, for before Newton we did not know that gravitation was a universal phenomenon (see Nelson 1908, §6). Thus for Aristotle the heavenly bodies did not gravitate. Nonetheless, in other contexts the more usual and shallow translation is necessary, for most commentators of Kant interpret his example blandly in the sense of, 'All (terrestrial) bodies are heavy'. See Chapter "Lecture VI" and especially Footnote 6 of that chapter for an example. The reader may incidentally notice that for Aristotle not even the bland statement was true, for he understood 'being heavy' as 'moving naturally towards the centre of the universe', and not all Aristotelian terrestrial bodies do so.

object the judgment is about. By contrast the truth of a synthetic judgment is never of the type that its negation entangles us in a logical contradiction. A synthetic proposition can be denied without any logical contradiction. If I say that a body does not gravitate, then this judgment is in itself as free of contradiction as the other —that it gravitates. To say that it does or that it does not is equally compatible with the concept of BODY that we find in Subject position.

It is therefore quite clear, on the basis of this simple distinction, that the law of non-contradiction is not an adequate criterion of truth, and that it is not just because of a limitation of the human spirit, as Leibniz maintained, that we cannot reduce all truth to this principle. The law of non-contradiction is, we might say, a sufficient criterion for the truth of analytic judgments. It only rules out those judgments that contradict themselves. What contradicts itself is logically impossible, but what does not contradict itself is not yet logically necessary, but merely logically possible. Hence, lack of contradiction is not a sufficient criterion of truth, but rather only a negative criterion of truth, as Kant says.[5] An assumption that goes against this principle is necessarily false, but one that does not go against this principle is not therefore true without further consideration. The criterion of the truth of synthetic judgments must therefore be sought outside the field of logic.

As clear and simple as this distinction between analytic and synthetic judgments is, and for all the clarity it offers concerning this question, some people find it hard and even cryptic.[6] To reassure ourselves that Kant's distinction is useful and appropriate I want to address briefly the most important objections that have been raised against it. Let us take Kant's example. That GRAVITY must be as much a universal and necessary property of bodies as extension is, has been argued for on the strength of the meaning of the proposition 'all bodies gravitate', viz that no body can lack the attribute of GRAVITY. This attribute belongs, so the argument goes, to the essence of bodies with the same universality and necessity as extension does. Is then 'all bodies gravitate' just as analytic as 'all bodies have extension'? Anyone who so argues is confusing the *concept* of BODY with the *essence* of a body. If by essence of a body we understand the set of properties that are common to all bodies and necessarily belong to each of them, then we must nonetheless distinguish the essence from the concept of BODY. For the concept of BODY in no way encompasses all properties that necessarily belong to each body, but only those that belong to the definition of a body, i.e. those properties which are necessary and sufficient to determine that something belongs to the class of bodies—and GRAVITY is not one of them. People had long known how to distinguish a body from what is not a body before they discovered that all bodies gravitate. Otherwise they could really have made that discovery through a fuller exploration of the concept of BODY. The discovery required in fact a completely different method. People needed to make

---

[5]See *Critique of Pure Reason*, A59 B84, A151 B190.

[6]The most notorious recent case is doubtless Quine (1951). His *arguments*, such as they are and as is usual in philosophy, cannot be said to have convinced everybody. For a recent and quite interesting metaphilosophical discussion of Quine's argument see Gutting (2009).

Lecture V

observations and experiments to establish universal gravitation, and that alone shows us that we are not concerned with an analytic judgment here.

This becomes even clearer if we consider that GRAVITY is not a predicate of a single body but rather a relation that a body exhibits in interaction with other bodies. Or let us take the example that I mentioned in my last lecture. The proposition that all radii of a circle are equal in length is an analytic judgment, since the mere definition of the concept of a CIRCLE as A CURVE ON A PLANE WITH A CONSTANT DISTANCE FROM A FIXED POINT is enough to see the truth of that proposition. On the other hand, the proposition that every circle is a closed curve also expresses a general property of circles, and is thus one that no circle can lack and that therefore belongs to the essence of a circle as much as the equality of the radii does. Yet this other proposition is no way analytic. If we assumed it to be false, i.e. if we assumed that a circle is *not* a closed curve, there would not be a *logical* contradiction. The assumption is quite compatible with the concept of CIRCLE according to the definition given. In order to see the truth of this other proposition we need a reason quite different from the mere concept of CIRCLE—we need truths originating in intuition.

What is instructive about this argument is that there are synthetic judgments that are as universal as the analytic ones, in other words that what we call the 'essence' of objects, is in no way sufficiently known through analytic judgments. This argument will be of use to us later on.

A second argument has been raised against this distinction. It is sometimes said that the boundary between analytic and synthetic judgments is not fixed, but instead ambiguous and indefinite, so that one and the same judgment can be analytic to one person and synthetic to another, indeed can even be synthetic and analytic for one and the same person at different times. This much is quite clear: should this view be correct, then there must be such a thing as a transformation of concepts. For whether a judgment is analytic or synthetic depends, according to the Kantian definition, only on what the concept in Subject position is. This concept would have to be capable of expansion if a judgment that was at first synthetic was later to become analytic. And some people have actually asserted this. A professor of philosophy whose lectures I heard as a student attacked the Kantian distinction between analytic and synthetic judgments by referring us to the example 'whales are mammals', which according to him showed how a synthetic judgment is transformed into an analytic one. For a zoologist who is reporting his discoveries about whales at a scientific conference, the judgment is analytic. In his concept of WHALE, as developed in zoology, whales are mammals. By contrast a city dweller or farmer who has never seen such an animal, and has always thought of a WHALE as a FISH, discovers something new when he learns that whales are mammals. For him this proposition is therefore synthetic. However, once he has been instructed by zoologists he will also have incorporated the attribute MAMMAL into his concept of WHALE. His concept has expanded and now the judgment 'whales are mammals' is also analytic to him. The same argument is used against the Kantian example. The judgment 'all bodies gravitate' was originally a synthetic one; one had to discover its truth; but now, knowing that GRAVITY belongs to the essence of a body just as much as EXTENSION does, the new attribute has been incorporated into the concept of BODY; it now belongs to it. The judgment is therefore analytic. And the same thing applies to the aforementioned example of the circle. If one comes to know that BEING A CLOSED CURVE is an attribute that belongs to the essence of CIRCLE as much as HAVING EQUAL RADII does, then both predicates have the same relation to the concept in Subject position. Both judgments are equally analytic.

What is the merit of this argument? There is indeed nothing against incorporating the attribute of GRAVITY into the concept of BODY, so that the sentence 'all bodies gravitate' expresses an analytic judgment, and the same applies to the other examples. But what has happened here? Has the concept expanded? Has a synthetic judgment turned into an analytic one? Not at all. What has expanded is our *knowledge* of the object in question. It is not the old concept that has expanded, but a new concept has been created, a concept that is of such a nature that if we start from it, then the same sentence that had earlier expressed a synthetic judgment would now express an analytic one. But mark my words: I am saying the same *sentence*. The Kantian classification, however, is not about sentences but about propositions or judgments. And the judgment which the sentence now expresses is different from the one it expressed earlier. We are only misled because the *words* are the same. The judgment expressed by the earlier sentence remains just as synthetic as it was before; we must only distinguish between two judgments that have the same verbal expression.

The important thing is this: that a term ('body', 'whale', 'circle') refers to the same *objects* before and after a change of its meaning, is a synthetic assertion, whose truth cannot be established by the fact that we are using the same *words*. On the contrary, we need reasons of a different kind. We must always bear in mind that two different judgments are expressed by one sentence. We associate two different concepts with one word 'body', and so we express two different judgments with one sentence. That a body according to the first definition is also a body according to the second definition (which includes GRAVITY as an attribute) is a synthetic judgment; and equally the judgment that a circle according to the first definition is also a circle according to the second definition is and remains a synthetic one.

References

Gutting, Gary. 2009. *What philosophers know: Case studies in recent analytic philosophy.* New York: Cambridge University Press.
Lambert, Karel. 2002. *Free logic: Selected essays.* Cambridge: University Press.
Nelson, Leonard. 1908. Über das sogenannte Erkenntnisproblem [On the so-called problem of knowledge]. *Abhandlungen der Fries'schen Schule (N.F.)* 2(4): 413–818. [Reprinted in Nelson (1971–1977), vol. II, pp. 59–393].
Nelson, Leonard. 1971–1977. *Gesammelte Schriften*, 9 vols. Edited by Paul Bernays, Willy Eichler, Arnold Gysin, Gustav Heckmann, Grete Henry-Hermann, Fritz von Hippel, Stephan Körner, Werner Kroebel, and Gerhard Weisser. Hamburg: Felix Meiner.
Quine, Willard van Orman. 1951. Two dogmas of empiricism. *The Philosophical Review* 60(1): 20–43.

Lecture VI

Abstract The distinction between analytic and synthetic has often been criticized by authors who have not understood and mastered it. A famous philosopher such as Schleiermacher and famous logicians such as Lotze, Sigwart and Trendelenburg were guilty of such lack of understanding but can be refuted by showing that they confused a concept with its extension, thereby falling into a kind of logicism.

We have got as far as discussing the distinction between analytic and synthetic judgments, this important discovery for which we have to thank Kant, and that casts so much light on the chaos that dominated philosophy before him. We might add that this chaos is still dominant as we speak, because the distinction remains an impenetrable puzzle for most philosophers. In fact, it is not even generally recognised as a discovery. The philosophers who profess this Kantian doctrine can be counted on the fingers of our hands—so small has its impact been thus far. This is an excellent example of how slowly any insight makes inroads in philosophy. Once a discovery has been made, it appears like the egg of Columbus, like a triviality, both to its discoverer and to anyone who understands it, yet everyone else considers it obscure, incomprehensible, wrongheaded, false or nonsensical.

In my last lecture I took care to point out the obstacles—at least the most important ones—that prevent us from understanding that rather simple distinction of Kant's. Without solving these difficulties it is not worth our while to delve into the finer and more subtle inquiries deriving from the Kantian distinction. But so that you may see that I am not tilting at windmills, I shall now read you some passages from famous recent logicians. They will document that some otherwise reputable thinkers find these difficulties insurmountable.

Our first case is Hermann Lotze, a famous German logician who taught several decades ago in Göttingen. When writing in his *Logic* about the division of judgments into analytic and synthetic, Lotze says *inter alia* the following[1]:

> In the general formula 'S is P' of the categorical [particular] judgment the generic concept S seems to be the subject, the generic P seems to be the predicate, and the constant, unchangeable and unrestricted union of S and P seems to be the meaning of the whole judgment. If we then explicitly add what is suggested (in any case intended) by the parallel

---

[1] See Lotze (1874, §57, pp. 79–80).

idea of [*S* being only] part [of *P*], then we will discover (1) that the subject is not really the generic *S* but rather an instance *Σ* of *S*; (2) that the predicate is not really the generic *P* but rather a particular modification *Π* of *P*; and (3) that the asserted relation does not really obtain between *S* and *P* but rather between *Σ* and *Π*, so that *this* relation (assuming the substitutions are right) is not synthetic anymore, indeed not even analytic, but rather a relation of identity.

According to the latter part of this passage Kant would be guilty of an incomplete division. In addition to analytic and synthetic judgments there would also be identical judgments. The identical judgments are of course an instance of analytic judgments. Indeed, whether the concept of the Subject is repeated partially or entirely in the Predicate is of no consequence for the question of whether by denying the proposition I produce a contradiction—which is the criterion for analytic judgments. Lotze then illustrates his general idea by means of an example[2]:

> We say 'Some men are black' and think we are hereby formulating a synthetic judgment, for the blackness *P* is not contained in the concept *S* of a man. Now the generic concept of MAN is not really the subject of this proposition (for it is not man who is black), but some particular men are the subject. By 'some', even though we *refer* to an indefinite part of mankind, we in no way *mean* such an indefinite part. For 'some men' are not here any men we may wish to extract from the set of all men. Our choice of 'some men' does not make them black if they are not black to begin with. We must therefore choose only those and *mean* only those who are black, in short the Negroes. (...) Again we say 'Dogs drink.' But the general dog does not drink; only one particular dog or several particular dogs or all particular dogs are the subject of this proposition. (...) Again 'Caesar crossed the Rubicon', yet not Caesar the baby but the Caesar who came from Gaul; not the sleeping Caesar but the one that was awake, conscious of the current political situation; not the hesitating Caesar but the one who had made his decision; in short the Caesar we mean by the subject of the judgment is the only one that the predicate defines—he who crossed the Rubicon.

You see that it all depends on a distinction which Lotze mentions without understanding its relevance for Kant's discovery—the distinction between the Subject in the sense of a generic concept in Subject position and the subject in the sense of the object of the judgment, i.e. the object which we judge to be black, to drink, and so on. That object has of course the attribute we predicate of it, at least if the judgment is true. If this is what we mean by 'subject', then all true propositions would be analytic. Inversely, analytic judgments would be true only because the object of our judgment is said to have a feature which the object in fact has. Yet the 'subject' in the sense of the object we are talking about in the judgment is something completely different from the Subject in the sense of the generic concept which is in the Subject position. And the relationship between the Predicate and this generic concept is alone relevant for Kant's discovery. Whether by 'some men' in the proposition 'some men are black' we mean those who are actually black (the minimum criterion of a true proposition) does not answer the question of whether it is an analytic judgment. The concept of MAN in Subject position does not contain all the attributes that belong to the object or the objects that are subsumed under that concept, but only those necessary and sufficient to define the object as an element of

---

[2] See Lotze (1874, §58, pp. 80–81).

Lecture VI

the class denoted by the concept. In order to determine whether something is or is not a man, I clearly do not need to know whether he is black. If by 'some men' I only mean black men, then consider the proposition 'Some men are albinos'—in continuation of Lotze's reasoning we would have to say that by 'some men' we mean albinos, and it would then follow that Negroes are albinos and albinos are Negroes.

Our second case is Christoph Sigwart, one of Lotze's successors in logic, and a no less reputed logician. When discussing the Kantian distinction, Sigwart takes the example 'This rose is yellow.' For him the subject of such a judgment is [not the concept but] 'this thing that is given to my intuition'; and what the judgment does is to analyse such a subject as follows[3]:

> One element of my intuition is identical with what I call yellow, and yellow is what I predicate of the whole [intuition] in my attributive judgment.

Sigwart means therefore that this judgment is based on an analysis of what is given in intuition. Even if we assume that Sigwart's account is perfectly right, it still is irrelevant to the analyticity of the judgment, contrary to what Sigwart thinks. To decide whether the judgment is analytic it does not matter whether yellow is one of the attributes I can discover through an analysis of my intuition of the rose. The only thing that matters is whether I can extract that attribute from the concept of ROSE. If so, then I would not need intuition to discover that the rose was yellow— there would be no rose of another colour, only yellow ones. Words like 'some', 'the', 'this', and so on indicate that we do not refer to the concept itself but to the object or to some objects that are subsumed under the concept. The concept ROSE is indeed no more yellow than the concept MAN is black.

Our third case is the logician Alfred Trendelenburg.[4] His example is 'This parabola intersects a circle.' After saying that 'this intuition', namely the intuition of the intersecting, 'does not lie in the generic concept' of the PARABOLA, he asks: 'But is the subject a generic concept?' Of course the subject [in the sense of the thing we are talking about] is not a concept, *this* parabola is not a concept, concepts do not intersect circles. This is certainly not what Kant meant. Of relevance is only that the predicate is not logically contained in the concept in Subject position. That's why Trendelenburg's example is doubtless a synthetic judgment.

Let us now tackle the other objection—Kant's distinction would not be precise, but one and the same judgment would be sometimes analytic and sometimes synthetic. Friedrich Schleiermacher, as far as I know, was the first to raise this objection. He says in his *Dialectics*[5]:

> The difference between analytic and synthetic judgments is fluid. The same judgment (e.g. 'Ice melts') can be analytic—viz if coming-to-be and passing-away through changes of

---

[3]See Sigwart (1889, §18, p. 137).

[4]See Trendelenburg (1862, vol. II, §XVI, p. 241).

[5]Schleiermacher's course and notes on dialectics were published posthumously in Jonas (1839). The passage quoted by Nelson is on p. 563 of that edition.

temperature was made part of the concept of ice—and otherwise synthetic. (...) The difference therefore just refers to the state of development of the concept.

Schleiermacher's example is ambiguous. The way he expresses himself misleads him. Consider the propositions 'Dog drinks', 'Rose is yellow'. We are usually more careful and avoid such misleading expressions lest it sound as though we actually intended to make a judgment about the concept. The concept ICE does not melt, only *all* ice or *this* [piece of] ice.

Allow me a digression before coming back to the ambiguity objection. After approvingly quoting Schleiermacher, Sigwart takes Kant's examples,

(1) All bodies are extended
(2) All bodies are heavy

and says about them the following[6]:

> Before having the experience that entitles me to the proposition, 'All bodies are heavy', I have only developed the concept of BODY by means of attributes such as BEING EXTENDED, and so on; but after having made the experience I can and must incorporate the attribute of WEIGHT into the concept of BODY in order to express the complete experience. My judgment, 'All bodies are heavy', is now an analytic one. Armed with this concept I could now proceed to have further experiences, e.g. 'All bodies are electric' or 'All bodies are warm'. If my concept would express complete knowledge (this would of course only be possible when we know everything about an object), then all judgments in the world would be analytic.

We are hereby led back to the most extreme logicism of Leibniz or his even more radical follower Wolff, for whom all judgments that express a sufficiently complete knowledge prove to be analytic—so much so that, according to Leibniz, God's mind must be capable of recognising the truth of all judgments just by using the law of non-contradiction.

The root of this fallacy was sufficiently clarified in my last lecture—the failure to recognise the difference between the concept of the object and our knowledge of the object. What expands is our knowledge of the object, but not our concept of it. The concept is unambiguously determined by the attributes that define it, and it cannot depend on the circumstances whether an attribute does or does not belong to the definition. There is as little ambiguity in this matter as there is in the disjunction 'a point either lies on a given straight line or it does not'. Changes in my knowledge do not affect the issue.

What is particularly misleading here is that nothing prevents us from defining a new concept and denoting it by the same word which we used to denote the original concept. What then changes is not the concept but the meaning of the word, i.e. the mapping relation between the one word and the concept denoted by it. As a result, the meaning of the sentence—i.e. the meaning of the linguistic expression (2)

---

[6]See Sigwart (1889, §18, pp. 134–135). This is an example of the bland interpretation of Kant's example; see Footnote 4 in Chapter "Lecture V" above. It is pretty obvious that Sigwart is not thinking of the momentous discovery of the law of universal gravitation, but only of the pedestrian observation that terrestrial bodies have weight.

Lecture VI 61

above—changes as well. Sentence (2) expresses both an analytic judgment and a synthetic one, depending on how we understand the word 'body', according to the first or the second definition. And there is nothing strange about this, for two different judgments can be such that one is analytic and the other synthetic. Yet what is most interesting about the above passage of Sigwart is what becomes explicit in the next bit. I shall consider it now, for it is instructive and takes us to the next point.

After incorporating the attribute of WEIGHT into the concept of BODY and thereby allegedly making the judgment analytic, Sigwart says[7] that he had no option, that he was able and forced to

> incorporate the attribute of WEIGHT into the concept of BODY in order to express the complete experience.

Thus, once I have experienced that all bodies, in the sense of the original definition, are heavy, I have to incorporate the attribute of WEIGHT into the concept of BODY as its defining attribute, in order fully to express my experience. In more precise words what one should say is that, if we want to have a concept of BODY such that it contains all attributes which we know a body to have and which a body in fact has, then we have to incorporate the attribute of WEIGHT into such a concept. Fine, but then we are not thereby expressing an experience, we are just constructing a new concept. What was our new experience about? It was about all bodies (in the original meaning of the word) being heavy, and *that* experience is certainly not expressed by a judgment in which the concept in Subject position, denoted by the word 'body', already contains WEIGHT. Such a judgment would indeed mean that not only everything that is extended, but in fact everything that is both extended and heavy, is heavy. *That* would be the explicated meaning of the sentence 'All bodies are heavy' if we should understand it in the sense of Sigwart's definition. So, if we incorporated the concept of WEIGHT into the concept of BODY, then the word form 'All bodies are heavy' means 'Everything that is extended *and heavy* is heavy.'

Does this express our experience? Can we at all experience that everything that is extended *and heavy* is heavy? No, we cannot experience *that*, we can only entertain it in thought, in fact this thought must be entertained because of the concept that is in Subject position. It is in fact an analytic judgment, and what an analytic judgment states can never be experienced. We cannot check through experience whether such a sentence is true or false. It escapes all empirical control. Take this piece of chalk. It certainly is a body. To subsume it under the new concept of BODY in Subject position implies asking whether chalk proves to be heavy in our experience or not. But to do that we must know beforehand that it is heavy, for otherwise we could not treat it as a subject and subsume it under the new concept of BODY. This we cannot learn from experience. It thus becomes clear that the truth is the exact opposite of what Sigwart has written. In order to express the whole experience we are not allowed to incorporate the concept of WEIGHT into the concept of BODY. If we insist on doing so, if we therefore conceive of a body as something

---

[7] See Sigwart (1889, §18, p. 134).

extended *and heavy*, then the empirical truth that we wanted to add to our scientific system is lost. In such a system the sentence 'Everything extended is heavy' (where the concept of BEING EXTENDED has been defined by means of the attribute of WEIGHT) is a new and analytic sentence, and the original one has disappeared. The only sentence that was supposed to express an empirical fact is then gone. All we have now is an analytic sentence that forever escapes empirical control.

In order to avoid any kind of misunderstanding, I will repeat that it is not forbidden to construct a new concept of BODY which besides the concept of BEING EXTENDED also has the attribute of WEIGHT. But it would be a grave error to believe that the old concept had thereby vanished off the face of the planet. It may indeed be ignored, or forgotten, and so may the empirical fact which it helped us discover; but it is not therefore wiped off. The old concept remains, and those who remember it, who wish to keep it alive and with it also the old judgment, for goodness' sake they can now proclaim it in sentences such as:

(3) All bodies are bodies
(4) All bodies in the sense of the first definition are also bodies in the sense of the second definition
(5) Everything extended that comes under the concept of BODY, in the sense of the first definition, also comes under the new concept of BODY.

These express now a fact of experience, namely that which we can only express by *not* incorporating the concept of WEIGHT into the concept of BODY in Subject position, then creating a new concept of BODY, so that uniting both concepts in a judgment is and forever remains a synthetic judgment. Just to be sure, I think I had better write down [on the blackboard] what the meaning of this seemingly trivial sentence (3) really is, viz.

(6) Everything that is extended is something that is extended and heavy.

Now *this* is the synthetic proposition that expresses the empirical fact that we had originally and simply expressed through the old sentence (2) 'All bodies are heavy.' The old synthetic proposition is and remains synthetic. You see this if you write it down explicitly. We can construct this new concept of BODY, of course. Any concept can be constructed as long as it does not contain a logical contradiction. But you can see that this new concept has a questionable property. Its definition is excessive, in that it contains more attributes than are necessary and sufficient to decide whether a given object belongs to the class. To define the class of bodies all we need is the concept of BEING EXTENDED. So, if our sentence, 'All bodies in the sense of being extended are heavy', is true, then I only need to know that the object in front of me is extended in order to infer that it is heavy, on the basis of our empirical proposition. The danger of over-defining concepts is that they easily make us forget that the relationship between the original concept, which had less content, and the new over-defined concept, constitutes a synthetic judgment.

There is another possible misunderstanding we should be aware of. A concept (like that of BODY) which is enriched by an attribute (such as WEIGHT) has the same extension as the old one, so that the objects subsumed under the latter also fall under the former. The extension of one concept is identical with that of the other. Yet asserting that the two extensions are identical is and remains a synthetic judgment. For we have to know an empirical fact in order to assert this identity. We cannot derive this from the concepts alone. We must therefore distinguish the identity of concepts (I mean, according to their contents) from the identity of their extensions. In my next lecture I shall go over a few applications of this distinction, so that you will appreciate how fertile and important it really is.

References

Jonas, Ludwig, ed. 1839. *Friedrich Schleiermacher: Dialektik*. Berlin: Reimer [This is vol. 2 of section 2 of the posthumous edition of Schleiermacher's works].
Lotze, Hermann. 1874. *Logik: drei Bücher vom Denken, vom Untersuchen und vom Erkennen*. Leipzig: Hirzel. [English translation *Logic in three books of thought, of investigation, and of knowledge*, ed. Bernard Bosanquet. Oxford: Clarendon Press, 1884].
Sigwart, Christoph. 1889. *Logik*, vol. I: *Die Lehre vom Urteil, vom Begriff und vom Schluss*, 2nd ed. Tübingen: Mohr. [English translation: *Logic*, vol. I: *The judgment, concept, and inference*. London: Swan Sonnenschein, 1895].
Trendelenburg, Adolf. 1862. *Logische Untersuchungen* [*Logical investigations*], vol. 2, 2nd ed. Leipzig: Hirzel.

Lecture VII

Abstract If everything that we find to be true of the objects subsumed under a given concept should be made part of that concept, so that all relevant judgments in which that concept appear would thereby be rendered analytic, then all knowledge of those objects would be destroyed. For synthetic judgments contain all the real knowledge we have about the objects of our inquiries, and from analytic judgments alone no synthetic judgment can follow. Analytic judgments have one logical function: to allow for inference without losing track of what we are talking about. This principle allowed Kant to reveal the different work done in geometry by definitions (analytic) and by axioms (synthetic).

My last lecture was about the objections raised against the Kantian division of judgments into analytic and synthetic. These objections, and the errors that result from them, play an important role in the study of philosophical fallacies. How important that role really is I shall endeavour to show you next, so please allow me to dwell a bit longer on Kant's distinction. The last question we discussed was whether something like complete knowledge of an object has to be expressed in analytic judgments, as most logicians keep proclaiming. I have tried to show that the exact opposite is the case, that the attempt to reduce synthetic judgments to analytic ones is in itself an impossible enterprise. In fact the correct thing to say is that restricting oneself to mere analytic judgments about an object results in erasing all real knowledge of it, all epistemic content, which is the only thing that makes any inquiry about the object worth our while.[1] This is so much the case that we might right away say that the ideal of those logicians, viz knowing everything about an object (so that we might derive everything from our concept of the object by explicating all that is implicit in it[2])

---

[1] The attentive reader will recognise the similarity of Nelson's difficulty with the epistemological conundrum so sharply formulated at the beginning of Frege (1892).
[2] This was also in fact Aristotle's conception of scientific knowledge. In the terms used by Nelson in Chapter "Lecture VI", which of course hark back to Aristotle, the definition of the concept has to be a complete statement of the essence of the object (for a recent study see Deslauriers 2007). Barnes (1969) explained that the Aristotelian conception was *didactic* in nature: once the research was finished, the definition would contain everything, so that the teacher would deploy the knowledge attained in such a way that the student might understand it (compare Hintikka 1972). The *reductio ad absurdum* that follows would then apply to Aristotle's conception as well.

would be the opposite of complete, and instead of knowing everything we would know nothing about the object. For anything we can know about the object would have disappeared.

To make this completely clear let us consider more closely an example I mentioned before—circles. A great many judgments can be made about circles. If we write [on the blackboard] those judgments as a sequence, we arrive at a numbered list of attributes of circles, e.g.

(1) The radii of a circle are all equal.
(2) A circle is a closed curve.
(3) A circle is a conic section defined by three points.
(4) A circle is a figure such that all inscribed angles that subtend the same arc are equal.

And so on and so forth.

Let us now imagine that we know everything about circles, that we are in complete possession of the attributes of this object. By following the definitional method just described we would now quite simply take the list of all those judgments on the blackboard as our definition of a circle. The concept of CIRCLE would be defined for us by the enumeration of all those attributes. This is so because the method prescribes that all attributes which we know to belong to circles become incorporated in the concept of CIRCLE, and the numbered list of attributes contained in a concept is nothing other than the definition of the class determined by that concept. I want to argue that whoever uses such a list of analytic judgments as his definition of a circle forsakes all knowledge about circles; for him all real epistemic content in this matter has disappeared. To convince yourselves of this just think that the word 'circle' in the above sentences *means* nothing else but an object that has all the attributes written on the blackboard. In order to recognise an object as being a circle we would have to know in advance that all those attributes belong to it, for only then can we subsume that object under the concept of CIRCLE.

What do we have when we get such a numbered list of analytic judgments? A tautology, a huge triviality, if you will a triviality that goes on and on ad infinitum. But lengthening the list does not make it less of a triviality. On the basis of this analytic judgment, of this set of analytic judgments about circles, we could never arrive at the conclusion that there is even such a thing as a circle. *That* would be a synthetic judgment, it would be saying something about the concept so defined which would go beyond the defining attributes. We would be saying that something falls under the concept, and in order to say such a thing we must go beyond the definition.

But let us ignore this argument, let us assume we know that something like a circle exists, in other words that this concept is not without an object. Even then we would not be able to say such simple things as that a figure whose radii are all equal 'is a closed curve' or 'is a conic section defined by three points' or 'is a figure such that its inscribed angles subtending the same arc are all equal'. By adding an existential statement to the effect that 'There are figures or there is one figure that not only has equal radii but is also a closed curve' I would know that; but whether

Lecture VII

all figures with equal radii are also closed curves is as undecidable as the inverse proposition [that all figures that are closed curves have equal radii]. And this applies to all relations between any two of the attributes mentioned in the above propositions, e.g. between (1) and (2), (1) and (3), (1) and (4), (2) and (4), and so on. Propositions connecting any of those attributes could not occur in science, even though they would constitute the whole epistemic content about circles.

I hope I have made clear that the allegedly complete knowledge that consists only of analytic judgments is in fact the most incomplete knowledge we can think of, for in it all real content of knowledge has disappeared and in its place there are just trivialities. The purpose of judging, as shown by the above arguments, always lies in a synthetic judgment, and analytic judgments would lack all import if they did not play a role as a means to this end. I shall now proceed to clarify what the import of analytic judgments actually is.

First then, it is only through their relationship with synthetic judgments that analytic ones have any import. If we start with a reasonable definition of the word 'circle', then we can say that all judgments written on the blackboard are synthetic, except for the first, which expresses the defining attribute. That first judgment is analytic, all the others synthetic. And so I argue that the import of such an analytic judgment is to enable us to make use of synthetic judgments. Their use is inferential, for we do not want to inquire every time we experience a figure whether it is a closed curve, has equal radii, and is a conic section defined by three points. In an inferential structure if we know this universal proposition that allows us to recognise a circle, then we can infer directly and without a specific examination of the particular figure, that it, being a circle, also has all remaining attributes. This is the whole point of an analytic judgment—to give us the criterion as to whether and when a given object may be subsumed under universal synthetic judgments. This criterion must be fulfilled to make the subsumption and all inferences therefrom.

Secondly, what kind of judgments are these that we infer from analytic judgments? My position is that they too are analytic, which helps explains the import of analytic judgments. There is no inferring without them. To make this clear we have to consider what is actually asserted at the end of a logical inference. A simple example, taken from the old textbooks, will suffice for the purpose:

All men are mortal
Socrates is a man
───────────────
Socrates is mortal

By the way, why is the word 'all' important in this simple inference? To draw the conclusion we need *universal* synthetic propositions, for only they license the subsumption we need. If only some circles were closed curves, then it would be of no use to me to know that a given object is a circle. I wouldn't know whether or not that object belongs to the part of the concept's extension where the closed curves are. The major premise of a logical inference must always be universal.

Back to our example. What does the logical inference say? It does not say that all men are mortal, nor that Socrates is a man, not even that Socrates is mortal. It only says that if all men are mortal, and if Socrates is a man, then Socrates is mortal. A logical inference consists in the derivation of an assertion from other assertions, or rather, and more cautiously, of a sentence from other sentences. To an inference naturally belong several judgments, the premises, and one conclusion. But neither the premises nor the conclusion are asserted in the inference, only the relationship between them, i.e. the relationship that we express by words such as 'it follows' or 'if—then' (in the sense of 'it follows'). There are also statements of the type 'if—then' that are not inferences, e.g. 'If a figure has all its radii equal, then it is a closed curve'. *That* is no logical inference, for the consequent does not follow from the antecedent, even though the proposition is correct. It is therefore a feature of a logical inference that the conclusion follows from the premises. We might say an inference is a hypothetical judgment, the 'if—then' expressing that the consequent follows from the antecedent, which just means that this hypothetical judgment must be analytic.[3]

You could argue that the *words* used by Kant in his definition[4] cannot be applied to such an analytic judgment. Kant's words refer indeed just to categorical judgments like the premises in the example ('All men are mortal', 'Socrates is a man') and its conclusion ('Socrates is mortal'). In other words, they refer to judgments that express a simple subsumption of an object under a generic concept. But the *meaning* of those words is such that that the division of judgments into analytic and synthetic can be easily understood to apply equally to hypothetical judgments and therefore also to logical inferences.

Consider what we say in the conclusion and what we say it about. We say it about what is asserted in the premises, and what we say is something new, viz that from those assertions follows the conclusion. If P is used to denote what the premises say, then what we say in the inference is that the conclusion follows from P. In other words, what is asserted by the premises (viz P) is the subject of the judgment which we call a logical inference; its predicate is 'from P the conclusion follows'; and the new judgment which asserts that the conclusion follows from the premises is an analytic judgment.

We say that it follows precisely because the premises give sufficient reason for the conclusion, i.e. we need no more reason than what the premises say in order to arrive at the conclusion. Kant's criterion for an analytic judgment is thus fulfilled, and so is

---

[3]Most examples given by Nelson of analytic judgments have the form of subject-predicate sentences, as were Kant's original ones. This paragraph makes clear, however, that he has in mind the concept of TAUTOLOGICAL, as introduced by Wittgenstein the very same year Nelson's course of lectures took place (see Wittgenstein 1921). Note that Nelson himself, a few paragraphs back, had used the same word, 'tautology', in a closely related context. The appropriation, for logical, or rather metalogical, technical purposes of this old rhetorical term (cf. Martin 1974, 301) must somehow have been in the air at the time.

[4]To wit, 'In all judgments in which the subject-predicate relation is thought..., that relation is possible in two ways. Either predicate B belongs to subject A as something that is (in a hidden manner) contained in concept A, or B lies completely outside of concept A, even though it *is* united with it. In the first case I call the judgment *analytic*, in the other *synthetic*.' See *Critique of Pure Reason* A6, B10.

the other one, viz that all analytic judgments are valid just on the strength of the law of non-contradiction. For indeed denying the conclusion would contradict the premises: if we assume for the sake of argument that Socrates is not mortal, then (a) upholding the first premise ('All men are mortal'), it would follow that Socrates is not a man, which contradicts the second premise; and (b) upholding the second premise ('Socrates is a man'), we would arrive at the proposition that not all men are mortal, which contradicts the first premise. Here lies the necessity of a logical inference: We cannot deny the conclusion without contradicting the premises.

We can clarify this result (viz that logical inferences are analytic judgments) by the sole consideration of the subjects of premises and conclusion. Let us take the general form of our example:

All M are P
S is M

S is P

We can formulate the inference as follows: A subject falling under the concept M, in such a way that all objects falling under M are P, is P. This sentence is clearly an analytic judgment.

Let us now look at some useful applications of Kant's distinction. Every logical inference and every proof (for proofs are simply chains of inferences) requires premises, some of which are needed as starting-points that make inferring possible and as such cannot be proved. Take a look at them; they constitute the primary presuppositions for all our inferences. I want to argue that if we wish to obtain a conclusion, a theorem, which is synthetic, and we want that under all circumstances (for it is only to this end that we infer and prove things), then there must be at least one synthetic judgment among our premises. If this condition is not fulfilled and we start only from analytic judgments, then no art of reasoning will ever allow us to derive a single synthetic judgment. In short, it is impossible to reduce a synthetic judgment to purely analytic ones.

The argument is straightforward. For let the opposite be the case, let a synthetic judgment be derivable solely from analytic judgments. In the inferential chain leading to the synthetic theorem, then, there must be at least one logical inference whose premises are both analytic and whose conclusion is synthetic. How would that be? Every conclusion requires a middle term M, i.e. a concept which appears in Subject position in the major premise and in predicate position in the minor premise. For it is by virtue of M that the Predicate in the major premise (the major term P) is applied to the Subject of the minor premise (the minor term S). Now our assumption was that both premises are analytic. So must the concept P be contained in the concept M and the concept M in the concept S, hence also the concept P in the concept S. Therefore, the conclusion would itself be an analytic judgment. It is therefore impossible ever to derive a synthetic judgment solely from analytic judgments.

The conclusion of this argument is a principle that refutes the way of doing philosophy we call logicism, which is what we started from. The endeavour of logicist philosophers amounts to deriving synthetic judgments from analytic ones alone. The impossibility of logicism has now been proved. The refutation of logicism is, however, relatively easy—the real task is to know how to apply the above principle to any particular case. Allow me then to give you some relevant indications by means of a few examples.

Our first example is a real classic, hugely important in the history of philosophy, for it led Kant to a great discovery. In Chapter "Lecture IV" we mentioned the geometric method, which Kant's predecessors pretended to use in order to upgrade philosophy into a science. In the course of criticising this method Kant discovered his distinction between analytic and synthetic judgments and this led him to a second great discovery. He first asked how the geometric method fares in philosophy. To answer it he compared philosophy and geometry, and asked what the effect of the geometric method in geometry is, and why geometers can so masterfully apply it and obtain so many results while all attempts to apply it in philosophy result in failure and produce nothing but confusion and more confusion.[5] This observation leads to a conjecture: There must be something at play in geometry [and absent in philosophy] besides the geometric method itself that explains why its application in geometry is so straightforward and productive.

And so it is indeed. For if we apply the division of judgments into analytic and synthetic to the propositions of geometry, then we first find that all definitions in geometry are analytic judgments, and we find secondly that all inferences [and proofs] in geometry are also analytic judgments. Inferences and proofs have the function of carrying over the certainty of the premises to the concluding propositions, i.e. they play the role of mediators. This is precisely the import of analytic judgments, as we have shown. Nonetheless, neither the definitions nor the inferences are the source of certainty, the reason why the propositions of geometry as such are certain. These certain propositions are not contained in the realm of mere definitions and inferences. The latter are only means to attain the former. So Kant's question is, 'Are the premises of geometry and therefore also their conclusions analytic or synthetic judgments?'

To ask the question is to answer it. The premises of geometry, the so-called axioms, are (as Kant says) all synthetic, and so are therefore all its theorems. The traditional observation that the proofs in geometry have to proceed in agreement with the law of non-contradiction had produced the fallacious conclusion that the propositions of geometry must be derivable from the law of non-contradiction. Geometry had always been the model of a science, which, according to Leibniz's distinction, belongs to the class of eternal and necessary truths. Everybody admitted that the truths of geometry are not factual, contingent truths, as Leibniz had designated the other class. Everybody knew that factual truths cannot be derived from the law of non-contradiction. Yet this proposition was inverted and people had

---

[5]These questions were already crucial for Kant's *Inquiry Concerning the Distinctness of the Principles of Natural Theology and Morality* of 1763 (see First Reflection, Walford 1992, 248–255). They occupy pride of place in the *Critique of Pure Reason*, A712–738 B740–766.

concluded that the truths of geometry, not being factual truths, must be of a purely logical nature, hence (in Kant's terminology) analytic judgments.

And so a deep fallacy has been uncovered which had dominated the entire history of philosophy until Kant—and we might add the entire history of philosophy after Kant as well, with extremely few exceptions, viz only excluding those few who made Kant's distinction their own. The next step is to combine it with Leibniz' division of judgments into empirical ones, i.e. those that assert contingent truths (Kant's a posteriori judgments) and those which assert eternal or necessary truths (Kant's a priori judgments). We see immediately that these two divisions do not cover the same ground. Each division is in itself complete. The synthetic judgment is defined through the negation of the attribute analytical and the a priori character of a judgment through the negation of empirical origins. Before Kant people had argued as follows:

> All *a posteriori* judgments are synthetic,
> therefore all *a priori* judgments are analytic.

And again:

> All analytic judgments (as shown in my last lecture) are a priori true and escape all empirical control,
> therefore all synthetic judgments are a posteriori.

There you have a typical philosophical fallacy, which has thus far dominated the entire history of philosophy.

A different question altogether is whether the propositions of geometry are a priori judgments or not, as assumed by Kant and all his predecessors. This question goes beyond my present task, so I shall not tackle it here. It only matters for our argument that, even if the propositions of geometry are found to be a priori, they are nonetheless synthetic. This argument raises the question of the nature of the axioms of geometry in the new terms made possible by Kant's distinction. The old question was easily answered: Are the axioms of geometry a posteriori? If they are not, then they are analytic. Are they analytic? If not, then they are a posteriori. The two distinctions were conflated and that conflation dominates contemporary philosophy. Such is the enormous significance of our argument—and not only for the history of philosophy.

References

Barnes, Jonathan. 1969. Aristotle's theory of demonstration. *Phronesis* 14(2): 123–152.
Deslauriers, Marguerite. 2007. *Aristotle on definition*. Leiden: Brill.
Frege, Gottlob. 1892. Über Sinn und Bedeutung. *Zeitschrift für Philosophie und philosophische Kritik* 100:25–50. [English translation: Sense and reference, *The Philosophical Review* 57(3):209–230, 1948].
Hintikka, Jaakko. 1972. On the ingredients of an Aristotelian science. *Noûs* 6(1): 55–69.

Martin, Josef. 1974. *Antike Rhetorik: Technik und Methode* [*Ancient Rhetoric: Technique and Method*]. Munich: Beck [=Handbuch der Altertumswissenschaft, II.3].

Walford, David., ed. and transl. 1992. *Immanuel Kant: theoretical philosophy, 1755–1770*. New York: Cambridge University Press.

Wittgenstein, Ludwig. 1921. Logisch-philosophische Abhandlung. *Annalen der Naturphilosophie* 14(3–4): 185–262. English translation: 1922. *Tractatus logico-philosophicus*, vol. 14. London: Kegan Paul.

Lecture VIII

Abstract Philosophers before Kant assumed that all a priori (non-empirical) judgments must be analytic, and also that all synthetic judgments must be a posteriori (empirical). Applied to the classical example of the axioms of geometry, these assumptions produced two opposing schools of thought—some philosophers said that geometry had to be empirical because its axioms were obviously synthetic; the other said that geometry had to be analytic because its axioms were obviously a priori. Kant's discovery that the two distinctions were not identical allowed for a middle ground position in which the axioms of geometry (as well as many other propositions) had to be synthetic a priori.

I have presented the discovery made by Kant thanks to his distinction between analytic and synthetic judgments, and I have started to highlight the importance and productivity of that discovery in relation to the main subject of this course. I beg you to keep in mind this connection, for what matters here is not so much the positive results of an inquiry (in this case Kant's inquiry about geometry) but the method followed, the questions asked, and the handling of the problem. For it is here that fallacies creep in and we must be on guard. And so I argued that the first question to be asked about a modern philosopher is whether he has made Kant's distinction his own. If not, then we need not concern ourselves with him any longer.

Let us now look into a second discovery made by Kant thanks to the same distinction, a discovery whose importance also lies in its setting of limits. Consider now the class of philosophers who have a grasp of Kant's distinction—do they also see that the propositions of geometry are synthetic? Again, if they do not, then there is no point in having a scholarly discussion with them, or this is what I want to argue in this lecture.

The best course to follow for this purpose is to start at the end—by considering briefly the conclusion I want to reach. It was fully clear (I hope) from my last lecture that in geometry, as in any other science, both definitions and inferences are analytic judgments. Hence, if we look at the content of synthetic propositions in a particular field of science, the first thing we have to do is put all definitions and inferences aside—they are not a matter for debate. Geometry is a case in point. If we find that the theorems of a science are certainly inferred from the principles of that science (in agreement with the logical law of non-contradiction) but are not on that account analytic, this issue being rather a matter to be decided by the nature of the premises, then all we need to examine are the first premises from which everything follows. If these too are all analytic, then no synthetic proposition can occur in the whole scientific field we are talking about, at least if we do honest work and play no tricks. So we must inquire into the axioms of geometry (for that is what we call the premises of geometry that cannot be reduced to mere definitions or what follows from definitions).

Take an axiom, whichever you want, and you will see that you are dealing with a synthetic judgment. This does not pose the least difficulty for anyone who has grasped the distinction between the two classes of judgments. Take the proposition used by Kant himself as an example[1]:

(1) The sum of the lengths of any two sides of a triangle is greater than the length of the third side.

In other words, every side is shorter than the sum of the other two sides. This proposition can be proved by the law of non-contradiction, for it can be reduced to the axiom which states that the straight line is the shortest distance between two points. This proposition is itself obviously synthetic, for the concept of STRAIGHT LINE does not contain any information about LENGTH, which is what the predicate is about. If we wish to define STRAIGHT LINE we might say that it is a LINE OF CONSTANT DIRECTION or something like that, but no definition will say anything about LENGTH. So if we talk about LENGTH in a proposition about STRAIGHT LINES, that proposition can only be synthetic. Consider another example:

(2) The sum of the three angles of a triangle is equal to two right angles.

There is nothing in the concept of TRIANGLE in (2) but the concept of a FIGURE GENERATED BY THREE INTERSECTING STRAIGHT LINES, i.e. there is nothing about the MAGNITUDE of the angles in that triangle or about their SUM. The proposition is therefore synthetic, just as every other axiom in geometry is.

Kant had thereby anticipated a highly significant, fruitful and profound discovery that came to preoccupy mathematicians in the nineteenth century. For anyone who knows something about the development of geometry in the last century it is certainly trivial and beyond question that the axioms of geometry are

---

[1] See *Critique of Pure Reason*, A25, B39.

Lecture VIII   75

synthetic propositions.[2] Yet this was not the case in Kant's times. People were then obsessed with the idea that geometry was the prototype of a science, science being about eternal truths, not truths of fact—in agreement with a distinction between two kinds of knowledge which (as you will remember) everyone took for granted at the time. To break with it was the sole merit of Kant's discovery, so that merely raising Kant's question about geometry was a tremendous audacity, a true revolution in thinking. He did not ask: are the principles of geometry empirical or logically necessary truths? (Under 'eternal truths' everyone thought only of those based on the law of non-contradiction, hence expressing a logical necessity.) Kant asked rather whether there might not be a third option, a kind of knowledge that has no place in the traditional division, viz judgments that are apodictically true—no empirical propositions—and yet synthetic.

I repeat that at this point I am not so much concerned with Kant's conclusion as I am with the *question* he raised, and I would like to dwell on it a little longer. If we use Kant's concepts to make the hitherto generally accepted division of judgments, then it amounts to a division of all judgments into analytic and empirical ones. Analytic judgments assert a logical necessity, whose truth depends on the law of non-contradiction, merely because of the concept of their object. Empirical judgments, on the other hand, are those where the connection of the concepts in the Subject and the Predicate position relies on an observation of the object itself, on a perception we need to have of the object, and so they are judgments that assert an empirical fact—a posteriori judgments, to use Kant's terms. Starting from these concepts, it is easy to see that we are dealing with a division that is in no way logically complete, whose completeness is not a purely logical affair. In other words, the following proposition,

(3) Every judgment is either analytic or a posteriori,

is certainly not an analytic proposition, but rather a synthetic one. It cannot be derived from the definition of these concepts. For if that were the case, the division would rest on a logical disjunction, in accordance with the principle of the excluded middle, and its more precise wording would be,

---

[2]This seems to have been the position of David Hilbert, who was Nelson's mentor in mathematics. It also was undoubtedly Frege's position, who, in contrast to Russell, did distinguish between the analytic truths of arithmetic and the synthetic truths of geometry (see Frege 1884). Even at the end of his life, when, as a consequence of Russell's paradox, Frege had given up the project of founding arithmetic on logic, he believed that arithmetic must be as synthetic as geometry (see Hermes et al. 1983). Poincaré's position is discussed later in Chapters "Lecture X" and "Lecture XI".

As we know, the logical empiricists and positivists, whose star had just begun to rise at the time this course was delivered, would defend the analytic character of geometry in terms that Nelson could no longer subject to scrutiny due to his early death. In any case, it may be opportune to say here that Nelson did not ignore or deny the existence of non-Euclidean geometries. In fact, he dedicated to them no less than two lengthy papers, one technical (Nelson 1905) and the other popular (Nelson 1906), as well as two oral expositions (Nelson 1914, 1928). See also Chapters "Lecture IX", "Lecture X" and "Lecture XI".

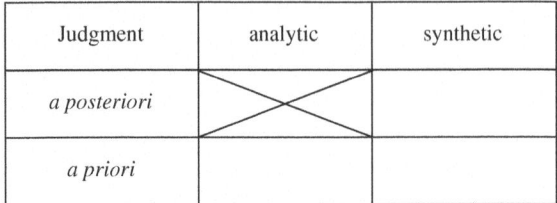

Fig. 1 Cross-classification of judgments according to Kant (analytic a posteriori judgments excluded)

(4) Every judgment is either analytic or not analytic.

And 'not analytic' equals 'synthetic', it does not equal 'empirical' nor 'a posteriori'. ANALYTIC versus SYNTHETIC—that is a logically complete disjunction.³ For the concept of SYNTHETIC is defined 'analytically' by the negation of the Predicate. This division is therefore certainly complete. The above division would also be logically complete if instead of 'analytical' we wrote 'non-empirical' or 'a priori'. For indeed the NON-EMPIRICAL JUDGMENT, the A PRIORI JUDGMENT, is defined as THAT WHICH IS NOT AN A POSTERIORI JUDGMENT. That would therefore also be a logically complete disjunction: A PRIORI versus A POSTERIORI. It does not refer to the content of the judgment, but rather to its origin, to whether or not the assertion that is expressed through it is based on the sense-perception of the object.

If we combine these two disjunctions, both of which are logically complete, then we arrive at more possibilities than were taken into account before Kant (see Fig. 1).

The two kinds of judgment that people before Kant thought possible were the analytic a priori and the synthetic a posteriori. But two cells are still empty. We have already argued in Chapter "Lecture VI" that there can be no analytic a posteriori judgments. The concept of an ANALYTIC A POSTERIORI JUDGMENT leads to a contradiction, for ANALYTIC JUDGMENTS are defined as those

JUDGMENTS WHOSE EPISTEMIC FOUNDATION IS CONTAINED IN THE CONCEPT THAT OCCUPIES THE SUBJECT POSITION,

and so are

TRUE INDEPENDENT OF ALL OBSERVATION OF THE OBJECT.

---

³The phrase 'complete disjunction' (and its opposite 'incomplete disjunction') refers in Nelson's German to a division of a concept (genus) into two sub-concepts (species) according to traditional logic, i.e. so that these two are mutually exclusive and jointly exhaust the superordinate concept. It is a very important term for Nelson's analysis in that the history of philosophy is marked by the presence of innumerable false dilemmas, all of which depend on a previous disjunction that appears to be complete yet is in fact incomplete. Figure 2 of this chapter, Fig. 1 in Chapter "Lecture X", Fig. 1 in Chapter "Lecture XIII", Fig. 1 in Chapter "Lecture XIX", Fig. 1 in Chapter "Lecture XXI" and Fig. 1 in Chapter "Lecture XXII" will give examples of several incomplete disjunctions and the false dilemmas that rest on them.

Hence analytic judgments are always a priori necessary, and this is itself an analytic proposition, and thus also an a priori judgment. The impossibility of analytic a posteriori judgments is itself an analytic judgment.[4]

Something completely different is the concept of a SYNTHETIC A PRIORI JUDGMENT. If we conflate those two divisions, as was commonly the case before Kant, then the concept of a SYNTHETIC A PRIORI JUDGMENT does indeed appear to imply a contradiction. That would of course be a judgment that was not analytic and yet not empirical. What facilitates this fallacious conclusion is the correct idea that an analytic a posteriori judgment is a contradiction in terms. Since the concept of an ANALYTIC JUDGMENT implies that it must be a priori, people presumed, naively and without further examination, that the inverse was also true, i.e. that all a priori judgments must be analytic. Or again: since the concept of an A POSTERIORI JUDGMENT implies that it must be synthetic, people concluded prematurely that the inverse was also true, i.e. that all synthetic judgments must be a posteriori, and considered this a logical necessity.

That an a priori judgment must be analytic and a synthetic judgment must be a posteriori were taken as analytic judgments, which prevented any further inquiry. Such an inquiry would of course make no sense if the question is already taken as decided on the strength of an analytic judgment. It would be as if one wanted to confirm analytic judgments on the basis of experience. An inquiry of that kind would make no sense—this much was known already by Kant's predecessors, so they did not even try to ask whether perhaps such judgments as the axioms of geometry might be synthetic. They considered it self-evident and necessary that, since they quite clearly do not express empirical facts, they must be analytic judgments.

I shall now read a sentence in which Hume expresses this with striking clarity. I choose Hume because he is an extreme empiricist and sceptic, who more than anyone was predisposed to doubt the logical necessity of geometric truths. Hume says[5] that mathematical propositions

> are discoverable by the mere operation of thought, without dependence on what is any where existent in the universe.

Hume's distinction between 'relations of ideas' and 'matters of fact' was a cornerstone of his theory of knowledge—the equivalent of Leibniz' division in English. Hume says next:

---

[4]The reader should be reminded that after Kripke (1980) we have become used to draw this table of judgments (or propositions) differently, i.e. combining the distinction between a priori and a posteriori with that between necessary and contingent (rather than with that between analytic and synthetic). Viewed in that way, it can be argued that none of the four cells is actually empty; in particular there may be necessary a posteriori propositions as well as contingent a priori ones. This certainly appears to be quite non-Kantian; but this is not the place to take up the issue.

[5]All quotations refer to Hume (1748, Sect. IV, Part I).

> Matters of fact, which are the second objects of human reason, are not ascertained in the same manner; nor is our evidence of their truth, however great, of a like nature with the foregoing. The contrary of every matter of fact is still possible; because it can never imply a contradiction...

Hume says quite sharply that it is possible to think the opposite of every matter of fact. The opposite of a matter of fact can never imply a contradiction. This defines one class of truths. The other class, by contrast, which contains eternal truths, is not like that at all. They are necessary because assuming their negation implies a contradiction. So they are analytic propositions.

This fallacy, a typical one, consists in a conceptual conflation—taking ANALYTIC and EMPIRICAL as contradictory predicates, each one equivalent to the negation of the other. We thus get the two equations:

ANALYTICAL (LOGICAL) = NON-EMPIRICAL (*A PRIORI*)

EMPIRICAL (*A POSTERIORI*) = NON-ANALYTICAL (NON-LOGICAL)

This assumption leads to seeing propositions which are synthetic (viz 'all synthetic judgments are a posteriori' and 'all a priori judgments are analytic') as analytic, indeed as logical identities. It is entirely correct that an analytic judgment is not empirical and equally true that an empirical judgment is not analytic; but these concepts, and this is the important thing here, are not defined by each other; in other words, one should not invert these propositions. A non-empirical judgment is not ipso facto analytic, and a non-analytical judgment is not ipso facto empirical. If these were definitions, then the propositions could be inverted, and if the inversion was permissible, then we would have a disjunction before us, the completeness of which would be logically true.

The fallacy can therefore be expressed by means of the following sentence:

(5) Every judgment is either analytic or empirical.

Proposition (5) is actually synthetic, and indeed not only synthetic but also (to put it mildly) extremely doubtful; yet because of the fallacy (5) appears to be an analytic judgment. The truth of (5) appears to be matter of pure logic, a consequence of the law of non-contradiction. Any attempt to doubt this principle seems to lead to a logical contradiction. How could it be any other way, if both these concepts are defined reciprocally, each one by the negation of the other? The fallacy consists in equivocating by taking the concepts of LOGICAL and EMPIRICAL and replacing them with different and arbitrary concepts that act as definitions of the original concepts. People believe they are defining LOGICAL by the negation of EMPIRICAL and the other way round. This is the basis of all the fallacious conclusions drawn from what is now revealed as a false synthetic proposition—and they dominate the present philosophical landscape to such an extent that if all those fallacious conclusions were revoked not much would remain of contemporary philosophy.

This I shall illustrate through examples of the way philosophy after Kant has reacted to his division of judgments. To repeat, I am not concerned now with the truth of Kant's answer (even though I think that answer *is* unquestionably true) but

Lecture VIII

only with the audacity of his question (I mean the question whether e.g. the axioms of geometry, albeit non-empirical, might after all be synthetic). Anyone who labours under the fallacy we have been discussing will be forced in the case of any single judgment to infer either that it is analytic because it is a priori or else that it is empirical because it is synthetic. And so we have in a sense two counterbalanced approaches in the way people do philosophy when considering specific judgments. Let us take any axiom of geometry, and it will have one curious feature—to be a judgment that conjoins the apparently contradictory attributes of being synthetic and a priori. The philosophers who see here a contradiction can be divided up into two feuding parties. The one party, having clearly in mind the synthetic character of the proposition, will unavoidably proclaim its empirical origin. The other party, in the full consciousness of its non-empirical character, and being in the grip of exactly the same fallacy, will declare it to be an analytic judgment. Each side does the judgment as much injustice as the other. Let us make this point clear by means of a diagram (see Fig. 2).[6]

The shared fallacy we have put on top: 'Every judgment is either logical or empirical.' I call this a fallacy[7] not because it is a false proposition, but rather because it is taken to be a logically complete disjunction—and this it is not. (I could just as well write one of the above-mentioned equations instead.)

Our example is geometry and its axioms. We have, on the one hand, the proposition, 'The axioms of geometry do not stem from experience.' This proposition, like the one I just read from Hume, asserts that they are a priori judgments. On the other hand, there is the proposition that Kant was the first to discover: 'The axioms of geometry do not stem from logic.' That is the peculiar and instructive nature of geometric axioms—they give us examples of propositions that have the

---

[6]Figure 2 contains the first of several diagrams of the kind Nelson used to draw in his lectures and publications. As noted in the introduction, these diagrams were highly praised by Popper at the time (see Popper 1979). They consist of *boxes* connected by *lines* (mathematically speaking, they are directed graphs).

Each *box* contains a proposition, usually expressed fully by means of a sentence, although in some cases the proposition is just indicated by a noun phrase; see Fig. 1 in Chapter "Lecture XIX" for an example. Each proposition is metalogical in character, i.e. it does not say things about the world but about our concepts of the world. As for Fig. 2 in this chapter, note that of the four propositions in the middle of the diagram the two on top directly contradict the diagonally opposite ones.

The *lines* represent a logical derivation that is always directed from top to bottom. For the purposes of this book I have replaced those lines by arrows to make Nelson's convention perfectly explicit. Incidentally, whenever two (or more) lines reach the same box, it is always understood that they are not independent premises, each one leading to the conclusion, but rather joint premises within one and the same inference. In spite of this obvious formal defect of his diagrams, Nelson was clearly a pioneer in argument mapping, a recent and currently flourishing area within the theory of argumentation (see e.g. Monk and van Gelder 2004; van Gelder 2005; Rowe et al. 2006). His diagrams usually contain six, seven or eight boxes, although on one occasion he uses up to ten boxes (so in Nelson 1917, 629).

[7]I remind the reader that the German word is *Denkfehler*, literally 'error in (or of) thinking'. My justification for translating it by 'fallacy' is given in Chapter "Introduction".

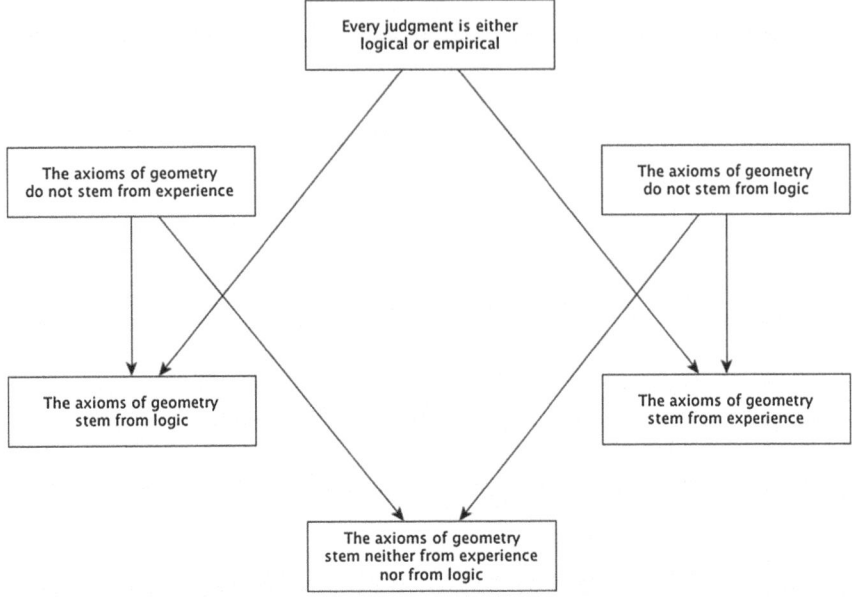

Fig. 2 Logicism versus empiricism in geometry: An incomplete disjunction

paradoxical quality of unifying these two attributes, or at least of making it clear that there is a problem in that a lot of propositions are like them.

Do you see the problem? If you presuppose that the axioms do not stem from experience, then the fallacy makes you infer that they stem from logic. On the other hand, if you presuppose that the axioms do not stem from logic, then the very same fallacy makes you infer that they stem from experience. So you are each time made to take one property of the judgments that you admit they cannot possess, and from that to infer the opposite property. These inferences are entirely equivalent. Saying that a thing which has the property A must have the property B is logically equivalent to saying that a thing that does not have the property B cannot have the property A.

In logic this is called contraposition. 'All As are Bs' is logically equivalent to 'All non-Bs are non-As'. Both sides are saying the same thing, only the starting-point is different, and so is the conclusion. The conclusion is always the negation of the premise of the opposite side. Kant started by discarding both conclusions as dogmatic, indeed entirely arbitrary assertions. Then, looking more closely, he argued that the proposition on top of the diagram is not only not a matter of logic but also in fact false. And so he inferred the existence of judgments that are neither logical nor empirical, that stem neither from experience nor from logic. In our next lecture we shall see how the philosophers after Kant reacted to this whole argument.

References

Frege, Gottlob. 1884. *Die Grundlagen der Arithmetik: Eine logisch-arithmetische Untersuchung über den Begriff der Zahl*. Wrocław: Wilhelm Koebner. [English translation by J.L. Austin: *The foundations of arithmetic: A logico-mathematical enquiry into the concept of number*. Oxford: Blackwell, 1950. New translation by Dale Jacquette: *The foundations of arithmetic*, New York: Pearson, 2007].

van Gelder, Tim. 2005. Teaching critical thinking: Some lessons from cognitive science. *College Teaching* 53(1): 41–46.

Hermes, Hans, Friedrich Kambartel, and Friedrich Kaulbach (eds.). 1983. *Gottlob Frege: Nachgelassene Schriften*. Hamburg: Felix Meiner.

Hume, David. 1748. *Philosophical essays concerning human understanding*. London: A. Millar [This is now widely known as *An enquiry concerning human understanding*, according to the second edition of 1772].

Kripke, Saul. 1980. *Naming and necessity*. Oxford: Blackwell.

Monk, Paul, and Tim van Gelder. 2004. *Enhancing our grasp of complex arguments*. Plenary address to the 2004 Fenner conference on the environment, Australian Academy of Science. [Revised version available at www.austhinkconsulting.com].

Nelson, Leonard. 1905. Bemerkungen über die nicht-euklidische Geometrie und den Ursprung der mathematischen Gewißheit [Remarks on non-Euclidean geometry and the origin of mathematical certainty]. *Abhandlungen der Fries'schen Schule (N.F.)* 1(3): 373–430 [Reprinted in Nelson (1971–1977), vol. III, 3–52].

Nelson, Leonard. 1906. Kant und die nicht-euklidische Geometrie [Kant and non-Euclidean Geometry]. *Das Weltall* 6(10–12): 147–155, 174–182, 187–193 [Reprinted in Nelson (1971–1977), vol. III, 53–94].

Nelson, Leonard. 1914. Des fondements de la géométrie [On the foundations of geometry]. Paper at the Paris conference in which the Société internationale de philosophie mathématique was founded [Reprinted with German translation in Nelson (1971–1977), vol. III, pp. 129–185].

Nelson, Leonard. 1917. *Kritik der praktischen Vernunft* [*Critique of practical reason*]. Göttingen: Vandenhoeck & Ruprecht [Reprinted in Nelson (1971–1977), vol. IV].

Nelson, Leonard. 1928. Kritische Philosophie und axiomatische Mathematik [Critical philosophy and axiomatic mathematics]. *Unterrichtsblätter für Mathematik und Naturwissenschaften*, 35 (4). [Reprinted in Nelson (1971–1977), vol. III, 187–220. The posthumous publication of this address was preceded by a prologue by David Hilbert and followed by Nelson's replies to objections by Richard Courant and Paul Bernays].

Nelson, Leonard. 1971–1977. *Gesammelte Schriften*, 9 vols, ed. Paul Bernays, Willy Eichler, Arnold Gysin, Gustav Heckmann, Grete Henry-Hermann, Fritz von Hippel, Stephan Körner, Werner Kroebel, and Gerhard Weisser. Hamburg: Felix Meiner.

Popper, Karl. 1979. *Die beiden Grundprobleme der Erkenntnistheorie*. Tübingen: Mohr. English translation: 2008. *The two fundamental problems of the theory of knowledge*. London: Routledge. [The original German text was written by Popper in the late 1920s or early 1930s].

Rowe, Glenn, Fabrizio Macagno, Chris Reed, and Douglas Walton. 2006. Araucaria as a tool for diagramming argument in teaching and studying philosophy. *Teaching Philosophy* 29(2): 111–124.

Lecture IX

Abstract Our knowledge of the epistemological status of the axioms of geometry suffered from a fallacy in which the given concepts of logic and of experience were surreptitiously replaced by made-up ones (and thus a synthetic judgment by an analytic one). This led to the assumption that all knowledge stems either from logic or from experience, yielding two opposing solutions to the problem of the axioms of geometry: they are either analytic (logical) or empirical. In spite of Kant's argument that they were neither, the old logicist position was still being defended by philosophers such as Hegel in the nineteenth century.

We shall continue with the argument started in my last lecture. Do not be afraid that I will stay focused on geometry for the rest of the term. Geometry is simply an instructive example, and we shall see how useful it is for application to philosophical questions. If one uses the example of geometry to think through the point, then one will attain a complete grasp of the problem, for there is no better example than geometry for the purpose. In geometry we know precisely what we are talking about. Consider again the diagram that I drew on the board last time (see Fig. 2 in Chapter "Lecture VIII").

The problem is that the two premises seem incompatible—they seem so to contradict each other that starting from each premise one has to infer the opposite of the other. Now we argued that both inferences are unobjectionable if LOGIC is defined as

THE SYSTEM OF THOSE JUDGMENTS NOT STEMMING FROM EXPERIENCE,

and EXPERIENCE as embracing

ALL KNOWLEDGE NOT STEMMING FROM LOGIC,

in other words if the following are logical identities:

LOGIC = KNOWLEDGE NOT STEMMING FROM EXPERIENCE, *ie* FROM OBSERVATION AND EXPERIMENT

EXPERIENCE = KNOWLEDGE NOT STEMMING FROM LOGIC, WHERE 'LOGIC' IS THE SET OF THOSE PROPOSITIONS WHOSE NEGATION IMPLIES A CONTRADICTION

If we posit these [negative] definitions, then the said inferences are necessary; but the intended meaning of 'logic' and 'experience' is altogether different. The above definitions do not license the use of any of the conclusions to refute the opposite thesis; yet this is what people are wont to do. That the axioms stem from LOGIC (defined as above) is used to infer that they cannot be synthetic, in other words that their negation implies a contradiction. Again, that the axioms stem from EXPERIENCE (defined as above) is used to infer that the opposite thesis is false, and so that the axioms come from observation and experiment. But look how the meanings have changed. LOGIC is now not defined negatively (in opposition to EXPERIENCE) but positively, as

> THE SET OF THOSE JUDGMENTS WHOSE NEGATION IMPLIES A CONTRADICTION.

In parallel fashion, EXPERIENCE is now not defined negatively (in opposition to LOGIC) but positively, as

> THE SET OF ALL KNOWLEDGE STEMMING FROM EXPERIMENT AND OBSERVATION.

That is of course the original meaning of these words, and that is the meaning used in the conclusions; but these conclusions have actually been smuggled in because the meanings have changed:

From a negative attribute —

NOT STEMMING FROM EXPERIENCE		NOT STEMMING FROM LOGIC

people infer a positive one —

ITS NEGATION IMPLIES A CONTRADICTION		IT MUST BE BASED ON EXPERIMENT AND OBSERVATION

So people take a negative concept (defined only by exclusion of the other concept) and replace it by a positive one.

In order to carry out this inference a tacit second premise is needed and that is none other than the bridging proposition which asserts that the positive and negative concepts are interchangeable. I made that premise explicit as: 'All knowledge stems either from experience or from logic.' This makes the two concepts interchangeable; only then do they constitute a complete disjunction, only then is it possible to use one of the premises to infer the falseness of the other. This second premise is not mentioned because people are not aware of it, and people are not aware of it because they consider it trivial, the distinction being thought of as a mere matter of words: the concepts are interchangeable because they have the same content. Concepts that have the same content *are* indeed interchangeable: this proposition need not be expressed, it is self-evident, a logical triviality. And yet: the proposition that one

Lecture IX

needs to infer in the way described is *not* a logical triviality, for both concepts *do* have a different content, and their interchangeability is not a matter of logic. It is rather a synthetic proposition that has to be independently argued for.

To have uncovered this situation, to have recognised that the above dilemma is a false one, to have broken with the traditional division of knowledge into logical and empirical—that is the Kantian discovery that has been occupying us, as confirmed by our examination of the axioms of geometry. These axioms are the clearest, the most straightforward, the most striking refutation of the traditional division. If a third possibility is in fact realised, then the disjunction cannot be logically complete. The actual realisation of such a possibility is not required to argue against the logical incompleteness of the division. A logical argument that there is a third possibility would be enough, and we do not need to raise the question whether it is realised as a matter of fact. But the fact that the third possibility is actually realised makes it easier to recognise the logical incompleteness of this disjunction. If something is actual, then it is all the more logically possible, and cannot imply contradiction. Kant demonstrates the *tertium datur* by means of the axioms of geometry as a result of an unprejudiced examination of the facts of the matter, after leaving aside all preconceived opinions and all supposedly established philosophical principles. It does not matter how difficult the two horns of the false dilemma are to reconcile; this is a concern to be considered later; it is of no importance now. What is important is to accept the facts as they are; and the facts indicate that the tacit premise is not only not logically fixed but actually false. I repeat that for the purposes of this course it does not matter that it is false. The only thing that matters is the insight that the disjunction, 'All knowledge stems either from experience or from logic', is incomplete.

All this is not just of historical interest for the assessment of pre-Kantian philosophy, but fruitful and of use, indeed indispensable, for our own time, or so I shall try to show you by way of an example that illustrates the fallacy I have uncovered, i.e. the swapping of concepts underlying the above propositions and arguments. I have already said that anyone who has been taken in by this prejudice necessarily draws a one-sided conclusion, depending on which of the two properties of the axioms of geometry he observes. Now, before and until Kant the logicist way of reasoning was the only game in town, but since then the scientific landscape has changed quite considerably. I mentioned before that the Kantian discovery had anticipated a very important issue that preoccupied mathematicians in the nineteenth century, viz the discovery that the axioms of geometry are synthetic and in this sense do not stem from logic. This implies that their negation does not imply a contradiction. Axioms are not logically necessary truths. This insight was still foreign to mathematicians in Kant's time, but after him it gradually came to dominate the field, at first in consequence of inquiries about particular axioms, and only much later as a general proposition.

The mathematical investigation led to a glaring confirmation of Kant's discovery through the actual demonstration of logically consistent geometries in which one or more axioms are replaced by its negation. This fact explains why after Kant both working mathematicians and those familiar with mathematical research were forced to turn their backs on the logicist way of reasoning. The rigid logic of this way of reasoning made them abandon the a priori character of the axioms, for it was no longer possible to doubt or play around with their synthetic character. So they concluded that the axioms stem from experience, i.e. that it was necessary to support them by experiment or observation.

Let us start with an example from within philosophy—a particularly famous philosopher, one moreover who today is positively celebrated, a philosopher whom some see as the pinnacle of the entire development of the history of philosophy. Let us start with Hegel. If we can show that the most famous minds have proceeded in the manner described, then we can conclude *a fortiori* that most other philosophers have as well. We shall need no more examples. In his *Lectures on Natural Philosophy* (posthumously published under this title) Hegel attacks the Kantian doctrine of the synthetic character of geometric axioms and indeed attempts a refutation on the basis of Kant's above-mentioned example concerning the axiom that the straight line is the shortest path between two points. Hegel is bold enough to put forward a *proof* of this axiom as follows[1]:

> Yet the given definition...

—mark: Hegel calls it a definition!—

> ... is clearly analytic, for the straight line can be reduced to the simplicity of direction, and quantitative simplicity yields the smallest amount, here the shortest path.

The quoted sentence has the considerable advantage of at least being intelligible. We are blessed in being able to grasp its meaning, not always the case with Hegel. And so we can tackle the task of determining whether it is true. Hegel is right to point out that the straight line in the said axiom is a line distinguished by the simplicity of its direction. Yet nothing is said about its size, there is no quantitative specification in that concept. 'And quantitative simplicity yields the smallest amount, here the shortest path.' This argument is interesting in more ways than one. To begin with, Hegel's argument says nothing about the straight line's path being *between two points*, although this is a crucial feature, for no mathematician will assert that every straight line is shorter than every curved one. The axiom only concerns lines between two points and of those it says that the straight line is the shortest path. We can therefore predict that Hegel's proof will be fallacious, since it does not make any use of that feature of the axiom. His proof would prove too much, for it would actually follow that every straight line is shorter than every curved one.

However, the main point is that Hegel's argument rests on a premise that is not quite explicit, viz 'and quantitative simplicity yields the smallest amount'. Let us make the premise completely explicit:

(1) What is simplest in relation to direction is also what is simplest in relation to quantity.

---

[1]See Michelet (1842, §256, p. 50).

Lecture IX

Should (1) be true, quantitative simplicity would follow from directional simplicity; or more generally:

(2) What is qualitatively simple must also be quantitatively simple.

First of all, proposition (2) is doubtless synthetic, hence quite certainly incapable of demonstrating that the axioms of geometry are analytic. For we know that if one needs a synthetic premise to prove a proposition, that proposition cannot be analytic (see Chapter "Lecture VII"). Hegel's refutation of Kant is already at fault here. Secondly, Hegel's premise is not only synthetic, and so unfit to the task, but in fact easily shown to be false. Oskar Schloemilch, a philosophically schooled mathematician, responded correctly to Hegel a long time ago, and his incisive answer deserves to be widely acknowledged. He said that if according to Hegel's axiom the qualitatively simplest must also be the quantitatively simplest, then one need only replace the words 'straight', 'curved', 'line' with the words 'white', 'coloured', 'jacket', in order to conclude that a white jacket is always shorter than a coloured one.[2]

A second example is a paper widely referred to by philosophers who still adhere to the logicist perspective concerning the propositions of geometry, since it purports to prove that Kant's thesis is false (so that the axioms of geometry are not synthetic). The paper was written by Robert von Zimmermann and published in the *Proceedings of the Imperial Academy of Sciences* of Vienna in 1871 under the rather aptly chosen title: 'On Kant's mathematical prejudice and its consequences'. The conclusion of his argument is this[3]:

> If mathematical judgments are not synthetic, then Kant's entire critique of reason loses all support.

The writer admits that if that Kantian proposition is conceded, then all consequences deriving from it are free from objections. But he argues that it is false and a consequence of Kant's mathematical prejudice. (A prejudice, by the way, shared by all mathematicians today.)

Zimmermann admits [in contradistinction to Hegel] that in the concept of STRAIGHT LINE there is only a qualitative attribute at play and no quantitative one. (The quantitative one is added by the predicate). But, says Zimmermann, Kant's argument about the synthetic character of this axiom suppresses the idea of a path *between two points*.[4] This adds to the concept of STRAIGHT LINE an idea repeated in the predicate, namely the property of BEING SHORTEST. These are identical ideas. Let us read the whole passage[5]:

---

[2] See Schloemilch (1877, 4–5, footnote).
[3] See Zimmermann (1871, 15).
[4] In other words, the axiom should read, 'A straight line *between two points* is the shortest path between those two points.'
[5] See Zimmermann (1871, 18).

> In this supplement, 'between two points', we have a quantitative feature and none other than the very same feature which is expressed by the predicate 'shortest'. The proposition is utterly analytic!

Let us assume for the sake of the argument that Zimmermann is right. As always in such cases we shall make progress if the *definiendum* is replaced by its *definiens*. We thus obtain the sentence,

(3) The straight line that is the shortest is the shortest.

There is no doubt that (3) is an analytic judgment. The question is whether (3) is what mathematicians—and Kant in their wake—mean when they speak about the axiom we are discussing. It is not. So the next question is whether *another* proposition is synthetic, say

(4) The straight line is the shortest line connecting two points.

Proposition (4) is in fact different from proposition (3), and proposition (4) is what we are talking about. In proposition (4) it is no longer possible to find a property within the concept in Subject position which would allow smuggling the predicate back into it. The ensnaring word 'between' does not occur. We can also write:

(5) The straight line that connects two points is the shortest (or in Zimmermann's words, 'is the shortest between both points').

My position is that (5), like (4), is quite certainly synthetic. It has the utmost practical importance. It can be used for predictions, for technical applications with the most far-reaching consequences, none of which would attach to an analytic proposition. I can, for example, use (5) to predict that if I want to travel on the shortest route between Berlin and Paris, I will definitely not be able to achieve this aim if I choose the route via Stockholm, a prediction that most certainly cannot be based on the law of non-contradiction.

If somebody wished to deny this, he would have to smuggle something about SHORTNESS into the phrase 'connecting line', as was done with the word 'between' above; he would have to say that the phrase 'connecting line' expresses a quantitative feature which is then repeated in the predicate 'shortest'—and the absurd consequences would become visible. If every line that connects two points were the shortest, it would follow that there was no such thing as a curved lines, or that every curved line is straight, or whatever other fanciful inferences might be confabulated.

One arrives at such absurdities if one attacks the synthetic character of geometric axioms. It is not worth dwelling on this logicist endeavour any longer. Let us turn our attention to the other perspective [viz that the axioms of geometry are empirical]. I do not want here to give an example from philosophers but rather from mathematicians who were well acquainted with the truth of the synthetic character of geometric axioms and who were the actual discoverers of that truth from the mathematical standpoint. If we can show that such excellent and acute mathematicians fell prey to the fallacy we are talking about, and indeed fall prey to it again and again, then we will think quite differently about the usefulness of our logical investigations.

References

Michelet, Carl Ludwig, ed. 1842. *Georg Wilhelm Friedrich Hegel's Vorlesungen über die Naturphilosophie*. Berlin: Duncker & Humblot. [English Trans. *Hegel's Philosophy of Nature*, 3 vols., London, George Allen & Unwin, 1970].

Schloemilch, Oskar. 1877. Philosophische Aphorismen eines Mathematikers [Philosophical Aphorisms of a Mathematician]. *Zeitschrift für Philosophie und philosophische Kritik*, 70: 1–15.

Zimmermann, Robert von. 1871. Ueber Kant's mathematisches Vorurtheil und dessen Folgen [On Kant's Mathematical Prejudice and its Consequences]. *Sitzungsberichte der Kaiserlichen Akademie der Wissenschaften, Philosophisch-Historische Classe* 27: 7–48.

Lecture X

Abstract The fallacy of concept swapping (i.e. replacement of a synthetic by an analytic judgment) which is responsible for the logicist position on the epistemological status of the axioms of geometry also underlies the empiricist position. The fallacy is shared by famous scientists (e.g. Schröder, Ostwald, and Mach), and it has pushed several high-calibre mathematicians (Gauss, Lobachevsky, Riemann, and Helmholtz) into empiricism, and another (Poincaré) into conventionalism. They all resisted Kant's solution—the idea of synthetic a priori judgments.

In my last lecture I explicated one form of a typical fallacy I want to make you aware of, leading to geometric logicism. Today I intend to clarify the opposite form of the same fallacy, leading to geometric empiricism, the view that geometry is based on experience, i.e. on experiment and observation. People arrive at this conclusion via a fallacious argument, as we made clear, viz the fallacious argument which starts from the synthetic character, the non-logical origin, of the axioms of geometry. You will recall the diagram that I drew on the board [Fig. 1 in Chapter "Lecture VIII"], and which depicted this relationship graphically. It is like a system of connecting pipes, where, if one decreases volume in one pipe, it rises in the other in equal proportion, and vice versa. The failure of geometric logicism is connected in the history of geometry with the appearance of geometric empiricism in its place. This empiricism is today's topic. Remember the shared fallacy of the two opposite conclusions in the diagram: the dogmatic disjunction of the sources of knowledge, the tacit presupposition that all knowledge must stem either from logic or from experience, a presupposition whose apparent self-evidence is the only reason for not making it explicit and so not discussing and examining it. This self-evidence is based on the swapping of concepts. One defines experience as that which is not logic, and logic correspondingly as that which is not experience. Let me write it down:

EXPERIENCE = NOT LOGIC = THE SET OF PROPOSITIONS THAT CANNOT BE SHOWN TO BE TRUE BY MEANS OF THE LAW OF NON-CONTRADICTION

LOGIC = NOT EXPERIENCE = NOT FROM INDUCTION (EXPERIMENT AND OBSERVATION)

Once the definitions are written down, we immediately recognise that the words 'experience' and 'logic' have two completely different meanings. For 'experience' is defined (1) as 'not logic', as not deriving from the law of non-contradiction, and (2) as stemming from induction, i.e. from experiment and observation; and no man will identify these two concepts:

STEMMING FROM THE LAW OF NON-CONTRADICTION

and

NOT STEMMING FROM EXPERIMENT AND OBSERVATION.

And yet it is precisely this false identification which our fallacy rests on. We can make the tacit presupposition explicit in logicist fashion so:

(1) Knowledge which does not stem from induction has to be based on the law of non-contradiction.

The empiricists base their reasoning in the corresponding pseudo-axiom,

(2) Knowledge that cannot be based on the law of non-contradiction has to be based on induction, and that means on experiment and observation.

We shall now look into proposition (2) a little more closely.

What kind of proposition is it? My thesis is that in the manner in which it is in fact used in the empiricist argumentation, proposition (2) is not logically necessary, it is rather a synthetic judgment. When we write it down explicitly, it becomes clear that (2) is itself not based on the law of non-contradiction. We can negate (2) without incurring the slightest self-contradiction. For who says that there are no other sources of knowledge besides experiment and observation and apart from the law of non-contradiction? There is absolutely no contradiction in assuming a third source of knowledge. The empiricist proposition (2) amounts to denying the possibility of synthetic a priori judgments. We can in fact formulate it just like that:

(3) Synthetic a priori judgments are impossible.

From proposition (3), and having demonstrated the synthetic character of a judgment, say of the axioms of geometry, it clearly follows that they cannot be a priori judgments but must be based on experiment and observation. Well and good, yet proposition (3) is firstly a synthetic judgment (as shown above) and secondly an a priori judgment. So it is a synthetic a priori judgment.

Why is (3) an a priori judgment? Kant's criterion for the a priori character of a judgment is strict universality and necessity. The word 'impossible' shows the criterion is fulfilled. Propositions that assert an impossibility can never be based on experiment and observation. Experiment and observation only ever refer to particular cases, to contingent facts, where we can at best, with greater or less probability, extrapolate to cases that have not yet been observed; yet we are never authorised to utter the word 'impossible'. Hence the empiricist proposition is

Lecture X

definitely a synthetic a priori judgment, and this is rather devastating for its own assertion—a synthetic a priori judgment that says no such judgments are possible!

I shall now show how widespread the empiricist proposition is and what a splendid reputation it enjoys, a task all the more convincing if I draw my examples from thinkers whose achievements in the field of exact science are beyond question. Among the most meritorious is Ernst Schröder, one of the founders of mathematical logic, whose *Lectures on the Algebra of Logic* (1890) are of course well known. This man, who certainly had a concept of what was required for scientific thought to be exact, stated the said presupposition unambiguously[1]:

> That sense perception is the original source of all knowledge is—now that the advocates of 'innate' knowledge have been defeated—only contested by those who believe in 'divine revelation'.

Let us be fair and assume that Schröder left analytic judgments unmentioned for simplicity's sake, not that he meant to say that analytic judgments (including the principle of identity and the law of non-contradiction) must be based on sense perception. So we shall only talk of synthetic judgments, for we have already seen that analytic judgments are a priori, indeed that this is itself an analytic judgment. So leave it aside. Now, despite the risk of being counted among the advocates of divine revelation, I question Schröder's proposition. We need only raise one question: How does anybody know that sense perception is the source of all knowledge? What is the origin and basis of such a universal assertion? Not sense perception certainly, for everybody admits that it relates only to particular, contingent facts and at best justifies assertions that extend to a finite number of cases, not an infinite one. Hence it is not possible to say something about all knowledge, as is the case here, without exceeding the authority of sense perception, i.e. of observation and experiment.

My second example is Wilhelm Ostwald, the well respected chemist and author of the *Lectures on Natural Philosophy* (1902a), who [in a different work] says[2] or rather posits as a principle of natural science that

> For today's natural scientist there is no a priori knowledge and therefore no apodictic knowledge. [...] In answer to Kant's main question: How are synthetic a priori judgments possible?, we answer: A priori judgments are not at all possible and all knowledge stems from experience.
> [...]
> Hence one can only assume a probability of $1/\infty = 0$ that any assertion that extends *ad infinitum* or is absolute hits the truth.

If anything is an assertion that is absolute or extends *ad infinitum*, then it is the assertion that a priori judgments are not at all possible, and since the probability that this assertion hits the truth equals zero, then the probability that Oswald's principle is true equals zero as well. It is therefore to all intents and purposes false. This

---

[1] See Schröder (1890, 3, footnote).
[2] See Ostwald (1902b, 51, 61).

empiricist argument is thus no better than the well-known sophism of the Cretan Liar. The Cretan Epimenides says, 'All Cretans are liars.' Fine, we assume this is true. All Cretans are liars, so what this Cretan says is a lie. His statement is therefore certainly false. The same thing happens to the above empiricist assertion.

Countless other authors talk like that and the same applies to all of them. I just mention Ernst Mach[3] on account of his being one of the most famous authors who defend the thesis that all knowledge stems from observation. We do not need any more names.

Let us now return to geometry, for the empiricist dogma also applies here. The following criticisms should not be understood as meaning that I want to diminish the achievements of the great men I am about to attack. What makes me do it is rather that some people have used the questionable doctrines of these mathematicians as authoritative enough to draw far-reaching philosophical conclusions from them. We must resist that. Dilettanti very often have the unfortunate bad habit of believing that one can add value to the achievements of great men by clinging to their fallacies. For their true discoveries are usually far less accessible to the non-expert.

A good example is Carl Friedrich Gauss, one of the first discoverers of the above-mentioned non-Euclidean geometries. Gauss had discovered the possibility of non-Euclidean geometry, i.e. he had discovered that it is possible to deny Euclid's parallel postulate—to deny in other words that through a point outside a given straight line in a given plane one can only draw one other straight line that does not intersect the first straight line—and to construct on that basis a consistent (contradiction-free) system of geometry. At the time this was an extremely paradoxical, surprising discovery, which greatly disconcerted Gauss himself because all his contemporaries believed that the axioms of geometry were analytic judgments, i.e. purely logical principles. Everybody believed the parallel postulate is a priori true, so it had to be analytic!

Confronted with the proof that one geometric axiom [viz the parallel postulate] was not logical in origin, the empiricist presupposition led people to believe that it was not a priori either. Gauss writes then that his discovery convinced him that geometry cannot be developed entirely a priori. 'Not entirely' were his words,[4] which suggests that Gauss only questioned the a priori character of that one axiom whose synthetic character he recognised, yet had no reason to doubt the a priori character of the other axioms.

Nikolai I. Lobachevsky, a contemporary of Gauss and co-discoverer of non-Euclidean geometry, argues in the same way. He says that, since the parallel postulate is not a necessary consequence of our spatial concepts, i.e. that it is a synthetic proposition, then only experience can confirm the truth of the parallel

---

[3]See e.g. Mach (1906, 314): "Let us now try... to analyse the course of research. Logic does not yield any new knowledge. Where does it come from then? It always stems from *observation*, which can be either 'external' and sensory or 'internal', concerning [our] representations."

[4]In a private letter to his friend, the mathematician Friedrich Wilhelm Bessel, dated 27 January, 1829. See Royal Prussian Academy of Sciences (1880, 490).

Lecture X

postulate,[5] and he proposes a way, viz to make empirical measurements of the three internal angles of a triangle. Lobachevsky did in fact invent methods of measurement of astronomic triangles in order to verify the truth of the axiom empirically. It is well known that Gauss undertook a measurement of this type on a terrestrial triangle, viz the one between the Brocken, the Inselsberg and the Hoher Hagen [three German mountains separated by 106, 84 and 69 km, respectively]. The result was that the sum of the angles does not deviate from the expected sum of 180° by more than the acceptable margin of error in observations.[6]

The third famous name in this field is Bernhard Riemann. He says[7] that, because

> the propositions of geometry cannot be derived from general quantitative concepts,

—and here see once more the sharpness and precision with which he expresses the presupposition of the synthetic character of the relevant judgments—, it follows that

> those properties which distinguish Euclidean space from other possible three-dimensional manifolds can only be derived from experience.

It is quite likely that Lobachevsky, along with Riemann, had never studied the relevant inquiries in Kant's *Critique of Pure Reason*. Not so Hermann von Helmholtz, who is most responsible for spreading geometric empiricism. Let us read a striking passage in which Helmholtz addresses this issue, albeit he is not here directly referring to the parallel postulate but rather to the axioms of congruence. Using the same methods of inquiry, Helmholtz had shown that these other axioms were equally synthetic, so that their negation does not lead to a contradiction. He says[8]:

> But if we wish to make inferences with logical necessity on the assumption that solid objects in space can move freely into every position in space without changing their shape,

—which is just an alternative way of formulating the axioms of congruence—,

> then we must raise the question whether this assumption may not imply a logically unproven presupposition. In the text that follows we will see that this assumption does imply such a presupposition, and indeed quite a consequential one. But in that case, every proof of congruence is based on a fact taken from mere experience.

Helmholtz clearly argues from the non-logical origin of axioms to their empirical origin, so that his argument undoubtedly presupposes, as its second premise, the pseudo-axiom we have been discussing all along. The most remarkable and historically interesting thing about this is that Helmholtz consciously directs his argument against Kant and seeks thereby to refute Kant's doctrine on geometry. His

---

[5]See Papadopoulos (2010, §9).
[6]For a popular account of these measurements see Kline (1967, 463–464).
[7]See Riemann (1867). The two phrases quoted by Nelson appear in the middle of the second sentence of that essay.
[8]See Helmholtz (1903a, 7).

refutation thus rests on a presupposition that was precisely the one contested by Kant. In other words, Helmholtz is begging the question.

Just how little Helmholtz was acquainted with Kant's previous thinking and writing on the matter emerges in a striking manner from another passage which has great interest for our argument. He applauds the English logician John Stuart Mill, whom I have mentioned before in passing, for having made a great logical discovery[9]:

> Mill was the first to make this important distinction between the properties that go into the definition of a concept (those which taken together are sufficient to fix a definition) and the properties that are additionally always present in particular objects that fall under the concept.

The one discovery that was the immortal achievement and the starting-point for Kant, viz the discovery of the distinction between analytic and synthetic judgments is here attributed to an English logician from the mid-nineteenth century.

An interesting development in the discussion of the origin of the axioms of geometry was introduced by Henri Poincaré, the French mathematician I mentioned before. Poincaré's great achievement (independently from Kant by the way) consists in forcefully arguing that geometric empiricism is untenable or at least that this solution to the puzzle of the non-logical character of geometry is not satisfactory. For, as he shows, its axioms are not propositions related to experimental research or to observation but instead are (and here, without realising it, he articulates a Kantian idiom) apodictic truths, and as such, i.e. as strictly universal and necessary propositions, they cannot stem from experience.[10]

Poincaré, however, confronted a conundrum, for he could not go back to geometric logicism in view of the well-established fact of the synthetic character of the propositions of geometry. The conundrum arose because he faced propositions that were evidently based neither on experience nor on logic. Anyone who inquires without prejudice, and out of pure love for truth simply starts from the facts, will nowhere find a more beautiful example than the manner in which Poincaré conducted himself in this impasse. He possessed the freedom from prejudice necessary to admit his conundrum openly—the impossibility of basing the geometric axioms on experience or logic. The only two sources of knowledge available having being drained dry, Poincaré's commitment to the truth went so far that in an act of desperation he came to the conclusion that the axioms of geometry are no judgments at all, given that all judgments stem either from experience or from logic. We thus need to change our former diagram as follows (see Fig. 1).

---

[9]Nelson quotes from memory; but although the passage is not literally faithful to the original, yet it does not betray its sense. See Helmholtz (1903b, §4): "Stuart Mill was the first to make this important distinction and so separated the properties that go into the definition of the concept (those which taken together are sufficient to fix a definition) from the properties that are additionally always present in particular objects that belong to the concept."

[10]See Poincaré (1902).

Lecture X

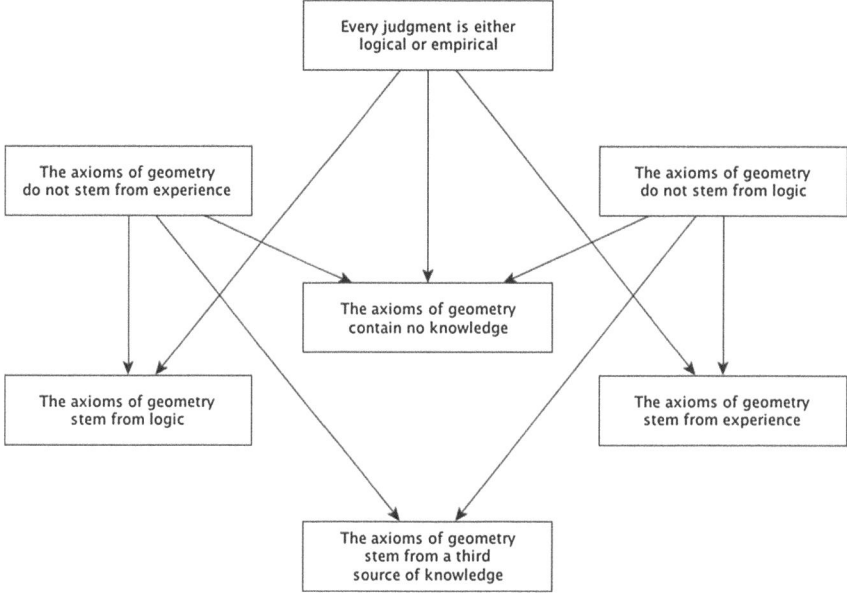

Fig. 1 Conventionalism: Poincaré's false way out of the logicist-empiricist dilemma in geometry

Poincaré started from the dogmatic disjunction of the sources of knowledge. Empiricism and logicism were blind alleys. He discovered that a further presupposition was at play here, a presupposition that, if it seemed to be somewhat trivial, was (he thought) nonetheless questionable, viz that the geometric axioms are in fact judgments and contain knowledge, knowledge of space or of spatial objects. This presupposition is shared by both horns of the dilemma, and his act of desperation was to eliminate this presupposition. By unifying the two correct presuppositions on both sides, and accepting the dogmatic disjunction of the sources of knowledge, he reached the conclusion: 'The geometric axioms do not contain knowledge.' They are then, in his words, mere conventions, arbitrary decisions. In this way he avoids both geometric logicism and geometric empiricism.

We can see that what forces this fatal conclusion is a presupposition that had not been noted before, not even by Poincaré [viz that all judgments stem either from logic or from experience]. If we only would drop that presupposition, we would find a way out of the conundrum. For we would then draw from the two opposing theses, the one on the left side of the diagram [viz that the axioms of geometry do not stem from logic] and the one on the right side [viz that the axioms of geometry do not stem from experience either], together with the other presupposition [viz that the axioms of geometry do contain knowledge] the conclusion that the axioms of geometry contain knowledge in the form of synthetic a priori judgments.

References

von Helmholtz, Hermann. 1903a. Über den Ursprung und die Bedeutung der geometrischen Axiome. In *Vorträge und Reden*, Fifth edition, vol. II, pp. 3–31. Braunschweig: Vieweg. [English translation: The Origin and Meaning of Geometrical Axioms, in Pesic 2007: 53–70].

von Helmholtz, Hermann. 1903b. *Vorlesungen über theoretische Physik* [A course on theoretical physics], vol. I. Leipzig: Johann Ambrosius Barth.

Kline, Morris. 1967. *Mathematics for the nonmathematician*. Reading: Addison-Wesley.

Mach, Ernst. 1906. *Erkenntnis und Irrtum: Skizzen zur Psychologie der Forschung*. Second edition. Leipzig: Johann Ambrosius Barth. [English translation: *Knowledge and Error: Sketches on the Psychology of Enquiry*, Dordrecht, Reidel, 1976].

Ostwald, Wilhelm. 1902a. *Vorlesungen über Naturphilosophie* [Lectures on natural philosophy]. Leipzig: Veit & Co.

Ostwald, Wilhelm. 1902b. Betrachtungen zu Kants Metaphysischen Anfangsgründen der Naturwissenschaft I: Die Vorrede [Reflections on Kant's Metaphysical Principles of Natural Science I: The Preface]. *Annalen der Naturphilosophie* 1: 50–61.

Papadopoulos, Athanase, ed. and transl. 2010. *Nikolai I. Lobachevsky: Pangeometry*. Zurich: European Mathematical Society.

Pesic, Peter (ed.). 2007. *Classic Papers from Riemann to Einstein*. Mineola: Dover.

Poincaré, Henri. 1902. *La science et l'hypothèse*. Paris: Flammarion. [English translation: *The Foundations of Science: Science and Hypothesis, The Value of Science, Science and Method*, New York, The Science Press, 1929, pp. 27–197].

Riemann, Bernhard. 1867. Über die Hypothesen, welche der Geometrie zugrundeliegen. *Abhandlungen der Königlichen Gesellschaft der Wissenschaften zu Göttingen* 13: 1–15. [English translation: On the Hypotheses that Lie at the Foundations of Geometry, in Pesic 2007: 23–40].

Royal Prussian Academy of Sciences, ed. 1880. *Briefwechsel zwischen Gauss und Bessel* [Correspondence between Gauss and Bessel]. Leipzig: Wilhelm Engelmann.

Schröder, Ernst. 1890. *Vorlesungen über die Algebra der Logik (Exacte Logik)* [A Course on the Algebra of Logic (Exact Logic)], vol. I. Leipzig: Teubner.

Lecture XI

Abstract Poincaré's conventionalism in geometry results from presupposing, as though this was a mere analytic judgment, that the sources of knowledge are only two—logic and experience—exclusive of each other and exhausting all the possibilities. That presupposition also led Einstein, via a misunderstanding of Hilbert's axiomatic approach, to the idea that the axioms of geometry are empirical in character. Our knowledge of the non-logical principles of mathematics in general and of geometry in particular is synthetic a priori, but has epistemological attributes that are lacking in philosophy. That explains why all attempts at using the mathematical method to attain philosophical truth are doomed to failure.

I had to be brief at the end of my last lecture as we had run out of time. We were talking about Henri Poincaré, and, as in the other examples we have examined, I would like to read you a passage written in his own words, so that you can form your own opinion. The book I want to read from, *La science et l'hypothèse* (1902), has been very poorly translated into German, so please do not trust the translation. This is really a shame, for the French original is a paragon of clarity, beauty, and terseness of expression. So I shall translate two passages for you[1]:

> Were the axioms of geometry, as Kant maintains, synthetic a priori judgments, they would then strike us with such force as to make it impossible for us to conceive their negation and to build a theoretical edifice upon that negation. There would be no such thing as non-Euclidean geometries.

Please listen again to the conclusion: 'They would then strike us with such force as to make it impossible for us to conceive their negation and to build a theoretical edifice upon that negation.' This would be so if the axioms were, as Kant maintains, synthetic a priori judgments.

However, the little word 'then' quite clearly cannot refer to the attribute 'synthetic'. Being synthetic does not by itself rule out the possibility of conceiving the negation of the axioms. If we were dealing with empirical propositions of experience, this would be immediately recognised—they would be synthetic judgments, only a posteriori. So the reason behind Poincaré's argument is the a priori character

---

[1] Poincaré (1902, 65).

of the axioms. The point is then again the idea that a judgment whose negation can be conceived cannot be a priori. So we have here again the much discussed dogmatic disjunction of the sources of knowledge in a new guise.

What has misled Poincaré and so many other thinkers are expressions such as 'necessary' or 'impossible to negate'. One would up to a point be justified in saying that the axioms of geometry, if a priori, are necessary in the sense that their negation, the possibility of negating them, is ruled out. But the question is why should it be ruled out? Is it because assuming the negation of an axiom contradicts itself? Or is it because it contradicts something we already know, e.g. our intuition of space or spatial objects, an intuition to which the axiom refers? Our inquiry therefore concerns this alternative: either axioms are impossible to negate because of an internal contradiction or because of a conflict with knowledge that we already possess from another source. In this case, no matter whether such knowledge be a priori or a posteriori, it is possible to conceive the negation of the geometric proposition without internal contradiction, and no more is required. To conceive the negation and to build a theoretical edifice upon such a negation is always possible in the case of synthetic propositions. The question of whether these synthetic judgments are a priori or a posteriori is neither here nor there. Hence a tacit dogmatic presupposition lies at the root of Poincaré's argument, viz that judgments a priori can only be analytic judgments.

The passage we have been discussing refers only to the question of the a priori character of axioms. Now listen to the second passage[2]:

> Where do the first principles of geometry originate? Are they imposed on us by logic? Lobachevsky has shown that this is not the case by his creation of non-Euclidean geometry... Do the axioms of geometry stem from experience? Deeper examination shows us that this is not the case.

We have here the two propositions, on the left and on the right of our diagram (see Fig. 1 in Chapter "Lecture X"), and the clear conclusion to be drawn from them:

> We will therefore infer that these principles are mere conventions.

The phrase 'are mere conventions' means 'cannot contain knowledge'. For if they did, then they would have to be either analytic judgments or synthetic a posteriori judgments. The first option (they are analytic judgments) is not tenable since Lobachevsky proved that the negation does not lead to contradictions. The second option (they are a posteriori judgments) is not tenable since they are apodictic, i.e. universal and necessary. And so Poincaré concludes that they cannot contain any knowledge at all. Once again the tacit presupposition, one that Poincaré is not aware of, becomes evident: All knowledge must stem either from logic or from experience. See again the diagram, which contains the interpretation introduced by Poincaré and his conventionalist argumentation. I dealt with the untenable consequences of conventionalism in Chapter "Lecture III". The same argument I used there against Poincaré's conventionalism in physics applies to his conventionalism in geometry.

---

[2]Poincaré (1902, 5).

Lecture XI

Our last example shows that the fallacy unfortunately plays an important role in the most recent exposition of the subject, the one given by Albert Einstein in his lecture on 'Geometry and Experience' for the Prussian Academy of Sciences, in which he discusses the question of the origin of the axioms of geometry. Einstein's main argument (not the only one, I hasten to add, for other secondary arguments accompany it incidentally, so to speak) follows the same kind of argumentation we have found in those other mathematicians I gave as examples. Einstein's starting-point is the new so-called axiomatic geometry. This expression refers to a particular way of doing geometry which inquires into the necessary and sufficient presuppositions needed to prove any particular geometric proposition. This method is optimal for the task of separating all those things in geometry that are logical in origin from all the other things that have a different origin. The result of such enquiries is all negative. For the only thing that can be established in this manner is that a proposition *cannot* be proved logically, hence it presupposes other sources of knowledge. Whether such sources lie in experience or elsewhere cannot at all be decided by this method. This is what Einstein overlooks when he infers the empirical origin of the postulates from their non-logical origin. He says[3]:

> According to axiomatic mathematics, only what is logically formal is an object of mathematics, and not the intuitive or other content associated with what is logically formal.

To be sure, axiomatic mathematics disregards all content that does not concern the logical relationship between propositions[4]:

> Such a purified exposition also makes it evident that mathematics as such cannot predicate anything about either objects given in our intuitive representation or about real objects. By 'point', 'straight line', and so on, axiomatic geometry gives us to understand only empty conceptual schemata. What gives them content does not belong to mathematics.

So far we can and must be in complete agreement, for Einstein is just expressing a negative result—that certain propositions are shown by this method not to have a logical origin. The misleading part is the special use of the word 'mathematics', meaning only axiomatic mathematics, i.e. the approach that only inquires into the logical relations between given propositions. Its questions are, in other words, whether or not a proposition logically follows from another one, whether or not a system of propositions is consistent, whether or not a proposition is necessary to prove another one. Questions like these are the sole concern of this kind of mathematics. If a proposition does not belong to this kind of mathematics, then nothing can be inferred from that proposition about its origin, except perhaps the negative conclusion that it does not stem from logic. Yet Einstein equivocates by taking the negative result and replacing it with the positive concept of EXPERIENCE, i.e.

---

[3] Einstein (1921, 4).
[4] Einstein (1921, 5).

STEMMING FROM EXPERIMENT AND OBSERVATION

or

STEMMING FROM INDUCTION

when he says[5]:

> Geometry thus purified...

—namely by [adding] the non-logical presuppositions needed to apply it—

> ... is clearly a natural science. We can consider it straightforwardly as the oldest part of physics. Its statements rely basically on induction from experience, not only on logical inferences.

In these passages everything that is negative is unassailable; only what is positive rests on a fallacious argumentation—'not only logical inferences, therefore induction from experience'.

What is the important lesson to be learned from all these examples? They acquaint us with the danger incurred by philosophers as well as by mathematicians, who know how to argue and how to think with rigour—that they may [unawares] return to that logicist scholasticism which is nowadays despised by everyone and from which empiricists in particular consider themselves to be worlds away. Faced with a factual question, viz whether or not an idea stems from experience, they fail to tackle it with methods appropriate to matters of fact but use instead a purely verbal argument. Combining the law of non-contradiction with a swapping of concepts made possible by an arbitrary definition, they present a synthetic judgment as though it was an analytic one (derived from their arbitrary definition). Thus the analytic judgment, 'Synthetic judgments stem from experience', where EXPERIENCE has been defined negatively as

THE SET OF ALL PROPOSITIONS NOT STEMMING FROM LOGIC,

is replaced by the synthetic proposition, 'Synthetic judgments are based on induction'.

There is another illuminating perspective on this fallacy. You will have noticed that the fallacious inferences are not positive and straightforward. They are negative arguments, polemically directed at one's opponent, the logicist against the empiricist and the empiricist against the logicist. Each rightly asserts that his opponent's view is false and then infers that his own positive view is true. These negative proofs have a tremendously corrupting effect in philosophy, for they are only fruitful when one has proved that between the assumption to be refuted and the assertion to be proved there is a complete disjunction [so that between the two they exhaust all possibilities]. Otherwise, one proposition cannot be inferred from the negation of the other. Philosophers overlook the need to prove that the presupposed disjunction is complete, and they avoid this task by letting the concept-swapping described do all the logical work. This is what happens with the disjunction

---

[5]Einstein (1921, 6).

EMPIRICAL versus LOGICAL. This disjunction would be logically impeccable if each member of this conceptual pair were defined as the negation of the other. But all we could infer from such definitions would be trivial negative propositions, which do not take us any further than expressing our disagreement with our opponent's standpoint without allowing us to establish our own position by means of a proper argument.

You might at this point say that critical analyses of the sort I have produced on those fallacies are rather unfruitful, for it would be much more to the point if we move on directly to a solution of the question itself. In all these lectures that question—'Are the axioms of geometry synthetic a priori judgments or not?'—has been postponed again and again, we have not answered it, and in fact we have not taken one step towards answering it. This may seem boring. But I trust you have seen the benefit of a discussion of real examples of objective attempts at a solution. All these attempts, no matter how great the men who put them forward, have proved to be fallacious arguments based on concept-swapping, because those men neglected to carry out a logical examination of the question before tackling it. To solve a question with some prospects of success, one has to be utterly clear about what the question means. We have to know what we are talking about, which precise assertion we are submitting for discussion, which other assertion we want to refute, what follows from one or the other. As long we are still unclear on this, it is futile to set about solving the objective difficulty, and this being the case there can be no hope of reaching a conclusion about the question of the origin of the axioms of geometry, regardless of all the hard work invested in it during the last century. More centuries will pass and the dispute will go on until people finally pull together and pose the antecedent logical question: what are we actually talking about?

How do the unsuccessful disputes surrounding that question in the nineteenth century and even today compare with Kant's simple, precise question: how are synthetic a priori judgments possible? By making the question precise, Kant was able, in one fell swoop, to anticipate the entire discussion that was to follow in the century to come. One need only compare a judgment such as the following, which I will read to you only because it is so typical, written as it was by the philosopher Friedrich Paulsen, a man who spent a large part of his life on the exposition, interpretation and clarification of this Kantian doctrine. He is talking of the distinction between analytic and synthetic judgments in Kant and arguing that the question about the possibility of synthetic a priori judgments is easily misunderstood[6]:

> One may say that it would have been better if [the question] had never been invented. Its impact on research has not been illumination but obfuscation. [...] The misleading formulation of the question of the critique may have contributed to so many post-Kantian works in the theory of knowledge missing the real problem which the dispute between rationalism and empiricism is about.

---

[6]Paulsen (1875, 171, 173).

That is the judgment of a man who dedicated a large part of his life to the task of clarifying this Kantian doctrine. How can we take offence if other people missed the meaning of Kant's question?

All of this was a digression occasioned by another Kantian question (see Chapter "Lecture VII"): why is the geometric method (as it was called at the time) so unfruitful when applied to philosophy? Why is it so effective in mathematics yet does not transfer to philosophy? Why have all attempts to do so failed and only increased the levels of dispute and confusion in philosophy? We can now quite easily show Kant's answer to this question, or rather we shall make the question somewhat more precise just as Kant did in order to answer it. The first question to ask is: how can we explain the success and fruitfulness of the so-called geometric method in geometry?

This question is in need of investigation, for we know that the essence of the so-called geometric method lies in nothing other than the rigour of logical inference, and if geometry should not possess another and more profound advantage than rigorous proofs, then the success of the geometric method in geometry would be as puzzling as its failure in philosophy. For logical inferences and proofs (which are only chains of such logical inferences), indeed the entire system of proofs—itself consisting of pure analytic judgments—is as unfruitful as purely analytic judgments can be. So there must be some other reason why using all those purely analytic-logical forms becomes useful and fertile. That reason is the epistemic content located in the axioms and the theorems derived from them. It is the unique nature of this knowledge that secures the success and fecundity of the geometric method in geometry, which for this very reason gave the method its name.

This follows from a simple comparison of the geometric axioms with the philosophical principles on whose basis the advocates of the geometric method in philosophy sought to establish a system. For while in geometry the most general principles from which the proofs proceed are in themselves clear and self-evident, the corresponding general principles of philosophy are altogether obscure and confused, indeed the more general they are the more confused and obscure they look.[7] In geometry the exact reverse is true. The most general propositions are the

---

[7]Descartes famously distinguished between 'clear' and 'distinct' ideas, although he never quite explained his distinction. An idea can be clear without being distinct but it cannot be distinct without being clear. What is added to *clarity* by *distinction*? Leibniz, in one of his few published philosophical papers, tried to do better than Descartes by proposing that an idea is clear if we can 'recognise the thing represented by it' and it is distinct if in addition we can 'enumerate one by one the marks which are sufficient to distinguish the thing from others' (Leibniz 1684; cf. Loemker 1969, 291–292). More tersely, we may say that if an idea is clear, then there is at least one thing we think by it; and if it is distinct, then not more than one. The opposite of 'clear' is 'obscure' (or 'opaque', 'foggy'), the opposite of 'distinct' is 'confused' (or 'blurred', 'fuzzy'). To say of a philosophical concept or principle that it is obscure means that we are unable to think *anything* by it; to say that it is confused means that we are unable to think *just one* thing by it. For a very different interpretation of Descartes's distinction, see Peirce (1878).

Lecture XI   105

most self-evident, and we can therefore proceed from them and confidently expect, if we keep our inferences rigorous, to arrive at ever new and certain results.[8] The lack of comparable self-evidence in the philosophical principles is the reason why all attempts to transfer the geometric method in philosophy have failed. This method consists in proceeding from the most general propositions and deriving everything else through inference. With this procedure one begins in geometry from what is most self-evident and clear, in philosophy from what is most obscure and confused.

From this a fruitful rule for the method of philosophy follows. In philosophy one can only hope to achieve fruitful results if one takes the reverse path, i.e. if instead of proceeding from what is most obscure and confused, one starts from what is clearest and most self-evident, particular truths that are not yet universal laws. I had occasion to mention this at the beginning of this course. The universal is most difficult to reach in philosophy, yet we are certain as to particular judgments, in judging particular cases. And we are only sure to arrive at the universal principle underlying the judgment of the particular case if we really start from the judgment of the specific case and then step by step ascend by abstraction from the particular content characteristic of the specific case to the universal principle, which is at first only an obscure basis for our judgment. What we need is thus a *regressive* method, one that ascends from the particular to the universal, from the consequences to their starting-points, and only then—having regressed to our final goal, having really completed the abstraction, having taken possession of the principles or highest premises, and only then—can the application of the geometric method serve a purpose in philosophy.

What I just said about the principles is equally true of the *concepts* used by mathematics or philosophy, and therefore also true of the role definitions play in both disciplines. To every definition we set forth belongs, if it is to be fruitful, a synthetic proposition that can never be derived from the definition, for only a synthetic proposition can inform us of whether the thing defined actually *exists* or whether the defined concept is without an object and we have only a mere fiction in our hands.[9] It was, for example, very easy for Leibniz to introduce the concept of

---

[8]This was probably true for the original axiomatisation of geometry by Euclid, who was apparently trying to realise Aristotle's ideal (see Heath 1949). However, it runs against today's consensus, according to which self-evidence and clarity are irrelevant to the selection of axioms (and definitions). It is interesting, and somewhat ironic, to note that Bertrand Russell gave in 1907 an interpretation of the rationale for an axiomatic system in *mathematics* that was quite similar to Nelson's idea of the regressive method in *philosophy* as expounded later on in this lecture, and in fact used the same name ('regressive') to refer to it. Given that Russell's paper was published posthumously and many decades later (see Russell 1973), it is very unlikely that Nelson knew about it. On the other hand, it is true that Russell mentioned the main point in passing at the end of the second paragraph of the Preface to *Principia Mathematica* (Whitehead and Russell 1910, v–vi, 1927, v).

[9]In spite of the differences in conception between Nelson, the defender of the analytic-synthetic distinction, and Quine, its arch-enemy, we see that there is agreement in one point: existential commitments are paramount when adjudicating between two theories.

his MONADS through a definition, more specifically the concept of a SLUMBERING MONAD defined as a SIMPLE SUBSTANCE WITH OBSCURE IDEAS.[10] Leibniz goes on to found his system of philosophy on this assumption. But his entire system rests on a mere fiction, for the existence of the thing he defines has not been proved. And what proof of existence could be given of the monad? From where would the knowledge come to construct such a proof?

Things are completely different in geometry. If the geometer defines, say, a CONE as a FIGURE GENERATED BY THE CIRCULAR MOVEMENT OF A RIGHT-ANGLED TRIANGLE AROUND ONE OF ITS CATHETI, then there is absolutely no lack of clarity or uncertainty over the question of whether we are perhaps dealing with a concept without an object, a mere fiction. For the definition contains a rule for the construction of such a figure, so that following this instruction convinces us of the existence of the defined concept.[11] The concept of a REGULAR POLYHEDRON WITH SEVENTEEN SIDES can also be formed quite easily by definition. But just as we were certain before of the existence of the defined object, so do we just as easily convince ourselves here that this is a mere fiction.

Now, if we wanted to emulate that geometric rigour in philosophy and sought to introduce new concepts by definition, we can now see where that leads. This is not just about avoiding introducing mere fictions; it is about something altogether more profound and of greater consequence. Philosophy is not in the business of defining concepts, for they pre-exist philosophy, and if we want to offer a definition for them, we do not only need to prove the existence of the thing defined (an endeavour likely to fail) but we must also prove that the thing defined is identical with what we conceived through the concept that existed before the definition.[12] If we neglect this task, the two concepts will get entangled and the properties that belong to the objects of one of the two concepts will be transferred to the objects of the other one. This is how concept-swapping enters the stage, as illustrated in the examples we have been discussing. I shall now leave these examples from geometry and proceed in my next lecture to examine analogous examples from philosophy.

---

[10] The example is taken from Kant, *Inquiry Concerning the Distinctness of the Principles of Natural Theology and Morality* of 1763, First Reflection, §1 (see Walford 1992, 249).

[11] This example corresponds to what mathematicians call a *real* definition as opposed to a merely *nominal* one. A real definition contains the instructions to construct the object, the nominal one just attaches a chain of words to a given term that might or might not have any actual reference. According to Nelson, as will become plain in the following lectures, philosophers abuse nominal definitions, fallaciously building their inferences upon them.

[12] This corresponds to the demand for *validity* in statistically oriented research.

References

Einstein, Albert. 1921. *Geometrie und Erfahrung*. Berlin: Julius Springer. [English translation: Geometry and experience, in Pesic 2007: 147–158].

Heath, Sir Thomas. 1949. *Mathematics in Aristotle*. Oxford: Clarendon Press.

Leibniz, Gottfried Wilhelm. 1684. Meditationes de cognitione, veritate et ideis. *Acta Eruditorum*, November. [English translation in Loemker 1969, 291–294].

Loemker, Leroy E., ed. and transl. 1969. *Leibniz: Philosophical papers and letters*, 2nd ed. Dordrecht: Kluwer.

Paulsen, Friedrich. 1875. *Versuch einer Entwicklungsgeschichte der Kantischen Erkenntnistheorie* [*An essay on the evolution of Kantian epistemology*]. Leipzig: Fues (Reisland).

Peirce, Charles S. 1878. Illustrations of the logic of science. Second paper: How to make our ideas clear. *The Popular Science Monthly* 12: 286–302.

Poincaré, Henri. 1902. *La science et l'hypothèse*. Paris: Flammarion. [English translation: *The foundations of science: Science and hypothesis, the value of science, science and method*, New York, The Science Press, 1929, 27–197].

Russell, Bertrand. 1973. The regressive method of discovering the premises of mathematics. In *Essays in analysis*, ed. Douglas Lackey, 272–283. London: George Allen & Unwin [Originally a paper read before the Cambridge Mathematical Club in 1907].

Walford, David, ed. and transl. 1992. *Immanuel kant: Theoretical philosophy, 1755–1770*. New York: Cambridge University Press.

Whitehead, Alfred N., and Bertrand Russell. 1910. *Principia mathematica*, vol. I. Cambridge: University Press.

Whitehead, Alfred N., and Bertrand Russell. 1927. *Principia mathematica*, vol. I, 2nd ed. Cambridge: University Press.

Lecture XII

Abstract We come back to the question why the logical method of rigorous logical inferences that works so well in geometry cannot be used in philosophy. Philosophical concepts are already in place before we begin philosophising, so that any attempt at defining them ends up in concept-swapping, i.e. replacing the original concept with a different and arbitrary one. Whenever philosophers do that, they equivocate. This fallacy is often compounded with circular definitions.

After a long digression I returned in my last lecture to our original question: why is it that imitating the method that has proved its mettle in geometry leads nowhere in philosophy? We obtained an answer by looking into the previous question: why is it that the said method proves so fruitful when applied to geometry? The traditional name 'geometric method', which expresses its aptness when we are after geometric truth, is neither here nor there. For the problem remains: why is it that the method of rigorous logical inference (starting with what is most general) works out fine when applied to geometric knowledge? *That* success is what lent it the name 'geometric method' in the first place. The answer came from observing that the axioms of geometry (i.e. the most general propositions from which geometry proceeds) were the clearest and most self-evident in the entire subject. Hence, the method by which one decides what is true in this or that particular case consists in reducing them to what is true in the more general case.

There is nothing analogous in philosophy; in fact we find the exact opposite. The most general propositions in philosophy are utterly obscure and confused, and we have no hope of ever attaining clarity and agreement except by reversing the direction of inquiry we were used to from mathematics. This entails two things. On the one hand, we have to take the most universal propositions, those which are in mathematics the starting-points, the principles that mathematicians consider most solid and secure, as it were the ground from which all questions can be decided, and make them into the very object of our inquiry. On the other hand, we have to take the particular truths, the individual empirical judgments upon which we are all agreed because of their very clarity, and make *those* our starting-points. And then we must take apart those starting-points, i.e. we must reverse the order of inquiry and look for their more general presuppositions, those that underlie them, even if obscurely. For only by a smooth transition that starts with what is maximally clear can we clarify step by step what is less clear. The clarification of what is in itself obscure consists in making clear that *it* is a tacit presupposition of what is clear to us. In other words, those things which are clear to us are true only if their obscurely underlying presuppositions are equally true.

This discovery, this bold reversal of the method familiar to us from geometry, and whose application Kant's predecessors had imagined would be capable of making philosophy into a science, is the great event, the turning point in the history of philosophy. For it makes it possible to raise philosophy to the status of a science, the feat mathematics achieved by the opposite method.

Let us once again state plainly what the decisive factor is here. We can express it by accepting a fact that pervades human knowledge, a fact that we must accept and submit to, the fact that the simpler and more general something is, the more difficult it is for us to grasp and the later it enters our consciousness. What is composite and particular enters our consciousness earlier than what is simpler and more general. Please do not mistake my meaning. When I say that in philosophy we have to start from the particularities of everyday experience and only from there to look for the transition to what is simpler and more general, we should not confuse such a regressive method with *induction*, which is also a regressive method, yet better known and used in natural science. Induction is also a transition from the particular to the general, from consequences to grounds. In induction, however, the transition takes place inferentially. The natural scientist infers from specific facts he has observed the general law on which they are based. The transition is, as we might say, one that proceeds from the consequences to their *real* grounds [i.e. to the reasons why something is the case]. In philosophy, however, one is concerned with the transition from the consequences to their *epistemic* grounds [i.e. to the reasons why someone knows that something is the case], to the grounds that are already logically presupposed in asserting the consequences and so quite certainly not derived from them in the first place.

So philosophy's regressive method should not be seen as an inferential procedure. It is rather a logical analysis of given assertions which are aimed at finding out what are the presuppositions entailed by this or that assertion. We need such an analysis in order to become aware of the presuppositions we do make as a matter of fact. Not until this regress from the particular to the general has been carried out, until by this regress we have step by step ascended to awareness of the most general presuppositions of our particular empirical judgments, is it time to build a system of philosophy, viz to take the general propositions we have discovered and make *them* the starting-point for a progressive inferential procedure.

A simple example of such a logical analysis, which will make clear that it is not an inductive inference, is the following. Anyone who is even a little adept in arithmetic and has just completed a lengthy addition will reverse the order of addition to check the result. If in reversing the order of addition he comes to the same result, then he has made sure that the original addition is correct. We will all naively use this method, but only a few are aware that in so doing they are actually presupposing a general law, and even fewer will be able to state the general law whose validity underlies the controlling procedure they have used. The general law of arithmetic,

$$a + b = b + a$$

is exhibited via the analysis of that common and generally accepted procedure as used in particular cases, and could only be discovered via such an analytic method.[1]

What I have just said of scientific propositions applies equally to the concepts used in them. Concepts are introduced in mathematics, and especially in geometry, by definition, using for this purpose some simple elementary concepts. The so defined concepts are again used to introduce further new concepts. Why can we not use this method of introducing concepts in philosophy? Note that the concepts which the philosopher has to use preexist any definition whereas in mathematics the opposite is the case, for mathematical concepts are formed by arbitrary definition in the first place. I recently gave the example of the CONE. This is a concept the geometer introduces by definition, one indeed which contains directions for constructing a cone. By following the directions the geometer can represent a cone in intuition. The use of this method in geometry prevents us from introducing concepts without an object or even contradictory concepts. For the required proof of existence can be more or less easily given, viz by using the axioms. In philosophy this procedure is unfeasible because, as I said before, we are already in possession of every concept before defining it, hence before the elucidation both of the elementary notions it presupposes and of the procedure by which they are to be put together so as to yield the concept. We obscurely presuppose philosophical concepts at all times when applying them to concrete cases. We use the concepts before we give an account (e.g. a definition) of them.

This can be very easily illustrated. Philosophical concepts used by all of us in everyday life in the most common empirical judgments are, for example, the concepts of TRUTH, EXISTENCE, POSSIBILITY, NECESSITY, CAUSATION, CAUSALITY, COMMUNITY, and so forth. We possess all these concepts before we even set about trying to introduce them by definition. This is even clearer in examples taken from the field of practical philosophy, in particular from ethics. Such are the concepts of VALUE and UNVALUE, MERIT and BLAME, DUTY, OBLIGATION, PENALTY, RIGHT, and others. Introducing one of these concepts by definition would require us to furnish proof of the existence of the thing so defined. Yet in philosophy we do not have an intuition we can use to display the object and dispel any doubt of its existence, in the way we display the possibility and existence of a cone in intuition by rotating a right-angled triangle around one of its catheti. The objects of philosophy lack this intuitiveness, they can only be *thought* of. No man has an intuition, or can hope to get an intuition, of NECESSITY, POSSIBILITY, NATURAL LAW, FORCE, CAUSALITY, DUTY, IDEALS, GUILT, and other such concepts, even although he can quite rightly exhibit objects of intuition of which he predicates those concepts, which in his judgment he subsumes under those

---

[1] Today we would say that the commutative law is provable by mathematical induction.

concepts. One can certainly point to an occurrence—therefore intuitively—and say of it, *this* is an effect, *that* is a cause, *this* is necessary, *that* is contingent, *this* action is a duty, *that* one is a crime. By intuition we know the object which we subsume under the concept in order to make such a judgment, yet we do not know by intuition *that* the object falls under that concept. *That* can only be *thought*.

If we therefore introduce a concept in philosophy by definition, the least we can say is that we have no means of convincing ourselves that the defined concept is not just a mere fantasy, a figment of the imagination, or even, what is more important, that it is inconsistent with the definition we have constructed. Let us assume that we manage both things, we convince ourselves both that the concept is consistent and that it has an object. Even in such a favourable case the procedure is not allowed, because the danger is always at hand of mistaking the concept as defined—call it D —for the original concept as given before the definition—call it C. In that case we would unconsciously transfer the properties of the objects that do fall under C to the objects that only fall under D. It is in any case an additional requirement to prove the identity of C and D. Only if and when such a proof is provided is a definition allowed in philosophy.

Such a proof can be provided by a procedure analogous to the one we recognised before as necessary for scientific propositions [e.g. the commutative law]. All we have to do is apply to concepts a corresponding method, viz a regressive one. We must thus take a given concept and analyse it with a view to those simpler concepts it already contains, and again analyse *these*, until we ascend to the simplest, unanalysable concepts. If we then survey which simpler concepts are contained in the given concept, we can go back and use them to define the concept originally given. And so the method of definition is allowed in philosophy if an analysis of the given concept is undertaken in advance which convinces us that the defined concept is really the same as the given concept. In philosophy we are thus not free to define as we are in geometry, but rather bound by the condition of explicating the concept beforehand. We can convince ourselves of the necessity of such a methodological reversal also in the case of concepts by considering examples. This is our next task.

An apt example to illustrate the misuse of definitions in philosophy is both simple and old—the arbitrary introduction of the concept of GENERAL OBJECT, well known for its important role in the philosophy of the Schoolmen.[2] But it has a corrupting effect even today among several philosophers who are not aware of this link. Locke's GENERAL TRIANGLE (of which nobody knows whether it is scalene, isosceles or right-angled, and so on) has gained notoriety, for it can be proved that it has both the one property and its negation, or no property at all, and even that it possesses neither the one nor the other.[3] What else would distinguish it from particular triangles?

---

[2] The term used by the Schoolmen was *ens universale, ens commune,* and occasionally *ens generale*. Behind all these terms lurks Aristotle's 'being *qua* being' (*Metaphysics* Γ).

[3] See Locke (1690, Book IV, Chap. VII, §9). Compare Berkeley's famous critique (1710, Introduction, §XIII).

Lecture XII

An example of the persistence to this day of the fallacious assumption of general objects is the concept of a GENERAL SELF (which plays a major role in the doctrines of many philosophers), or of CONSCIOUSNESS IN GENERAL, as other philosophers call it in contrast to the particular consciousness of individuals. People are easily led to the idea of introducing concepts such as GENERAL SELF, CONSCIOUSNESS IN GENERAL, or EPISTEMOLOGICAL SUBJECT (to mention a third term with the same function) by the following argument. Knowledge is always knowledge of an object and thus different from that object. If I know something, then what I know is not my knowledge and my knowledge is not what I know. The application of this proposition becomes difficult in the case of self-knowledge, the knowledge of our own self. It seems in this case necessary to assume that for knowledge of myself to be possible another self is required as the knowing subject, for according to that presupposition the knowing and the known, the epistemic subject and the epistemic object, must be different from each other. For a particular self to be a possible object of knowledge presupposes a general self, a non-individual self, a 'consciousness in general', in contrast to individual consciousness. How else should the individual self become an object of knowledge? But one knows of it, so there must be (one argues) a general self; otherwise no such knowledge would be possible.[4]

It is easy for a thinker schooled in logic to uncover the contradiction in this peculiar concept. It belongs to a family of contradictions first given a poignant, paradoxical form by the English logician Bertrand Russell, who is familiar to mathematicians through set theory. I will show that this paradox applies to the apparently innocuous philosophical concept of EPISTEMOLOGICAL SUBJECT. This is defined as

THE SUBJECT THAT KNOWS ALL SUBJECTS WHO DO NOT KNOW THEMSELVES.

Russell's well-known example of

THE SET OF ALL SETS THAT DO NOT CONTAIN THEMSELVES

is as contradictory as the SUBJECT THAT KNOWS ALL SUBJECTS WHO DO NOT KNOW THEMSELVES. For if such a subject knows itself, then it belongs to the subjects it knows, i.e. to the subjects who do not know themselves; therefore it does not know itself. But if it does not know itself, then it belongs to the subjects whom it knows; and so it does know itself.[5]

Another example (to a certain extent the opposite of the concepts that are reified as general objects) are the so-called fluid concepts that have just recently begun to

---

[4]See Nelson (1908, Chap. VI).

[5]The SET OF ALL SETS THAT DO NOT CONTAIN THEMSELVES was introduced in the context of Russell's famous contradiction or paradox (see Russell 1903, Chap. X). Nelson worked on the general logical form covering Russell's as well as other set-theoretic contradictions (see Grelling and Nelson 1908).

play a significant role in philosophy.⁶ Such concepts are supposed to be distinguished by the fact that they are not rigid, not determined once and for all, but instead adapt and accommodate themselves to all transformations of the objects they denote. People arrive at this notion of FLUID CONCEPT because they want to introduce an epistemic instrument capable of accommodating the change and flow of the facts of observation. They believe that the use of rigid concepts would impose the plainly incorrect assumption that the objects defined by rigid concepts must be equally rigid and unchanging. We have already argued against this and thus against the possibility of such fluid concepts when we asked ourselves whether a synthetic judgment can be transformed into an analytic one. This could not be done because concepts *are* rigid and cannot be transformed or expanded.

The fallacy in the FLUID CONCEPT case is similar to the EPISTEMOLOGICAL SUBJECT case. It is a fallacy of presupposition, which I shall now exhibit as such for both cases. From the proposition,

(1) Knowledge and object are necessarily different from each other

one draws in the EPISTEMOLOGICAL SUBJECT case the consequence that

(2) Knowing subject and known object must be different from each other.

But that is a *quaternio terminorum*, for even if all knowledge is different from its object, it does not follow that the knowing subject is necessarily different from its object, and so there is absolutely no contradiction in the possibility of self-knowledge. To be possible, the knowledge of the individual self does not need to assume a 'supra-individual self'.

Correspondingly, an incorrect inference is at play in the FLUID CONCEPT case, viz

(3) If this concept is rigid and it applies to that object, then that object is also rigid.

This argument confuses (4) with (5):

(4) This object falls under that concept.
(5) This object has the properties of that concept.

The best counter-example is the very concept that allows us to think of change, viz the concept of MOVEMENT. By subsuming something under this concept, we

---

⁶Nelson has Bergson (1903) in mind here. On p. 9 Bergson says that metaphysics only deserves its name 'when it overcomes concepts, or at least when it frees itself from stiff and ready-made concepts in order to create concepts quite different from the usual ones, I mean flexible, loose, almost fluid, always ready to mould themselves upon the elusive shapes of intuition'. For a more thorough treatment of this view, see Chapter "Lecture XIII". Bencivenga (2000) develops a very interesting argument to the effect that there has been a struggle between two kinds of logic going on ever since Hegel modified Kant's philosophy and consisting precisely in the acceptance or rejection of 'fluid concepts'. The reader might be aware of a huge literature on this very topic starting with Wittgenstein (1953) in philosophy and Rosch (1973) in cognitive science. This is no place to tackle what is a rather complex issue, but it may be interesting for the reader to consider Bencivenga's proposal jointly with the logical diagnosis Nelson presents in his Chapter "Lecture XXII", when he distinguishes between an Aristotelian-Kantian and a Neoplatonic-Fichtean logic.

assert that the object moves. We therefore quite certainly do not assert IMMUTABILITY, RIGIDITY, in this case IMMOBILITY of the object but rather the exact opposite.

Related to these examples is another that has a nefarious effect on contemporary philosophy—the concept of INDIVIDUAL CAUSALITY or of INDIVIDUAL LAW, an expression used by some philosophers that makes plain the contradiction in a straightforward way.[7] The idea is that the individual, especially in history, cannot be subject to laws, at least not natural laws. All natural laws are universally valid; this is part of their concept. Now the UNIVERSALITY or UNIVERSAL VALIDITY of a law seems to contradict the UNIQUENESS or INDIVIDUALITY of an historical fact. There seems to be a contradiction in saying that an historically existent individuality in all its uniqueness and unrepeatability is subject to general laws. Hence the idea of introducing the concept of INDIVIDUAL CAUSALITY. In order to understand historical phenomena in their mutual connections the concept of CAUSALITY is required—one phenomenon is to be understood as the effect of another. But to do this without contradicting the INDIVIDUALITY of historical phenomena, then the concept of INDIVIDUAL CAUSALITY must apparently be postulated. For the UNIVERSAL CAUSALITY of nature's laws (so the argument goes) would contradict the UNIQUENESS of historical phenomena—it would require those phenomena to be repeated again and again in agreement with the UNIVERSAL VALIDITY of the laws of nature. And so it comes that some philosophers take refuge in INDIVIDUAL CAUSALITY, a causal connection of a unique phenomenon with another unique phenomenon, that connection being also something unique, unrepeatable, individual.

So we have here again a logical contradiction. If I assert the causal connection of phenomenon A with phenomenon B, then I assert that if A takes place then B must also take place, and this hypothetical form (the only one in which we can give causal connections propositional form[8]) expresses the universality of a law of nature, a presupposition we have to make, even if it is an obscure one. There is no way I can assert the connection between phenomenon A and phenomenon B without thereby assuming that whenever A takes place B must also take place, i.e. without presupposing the validity of a law. The fallacy, the dialectic illusion that brings about the confusion, lies in neglecting that our judgment of a connection has the hypothetical form. For if we heed its hypothetical form, then it also becomes clear what it is that those philosophers are overlooking, i.e. that the general law in no way demands the repeatability of one or another phenomenon. For it does not assert that A takes place all the time nor that B does; it only asserts the connection between A and B as a law that is always valid. How often one or other phenomenon takes

---

[7]Examples of such philosophers will be given in Chapter "Lecture XIII".

[8]In Kant's original scheme the logical form of the conditional corresponds to the category of causality. Today this idea has been modified by having the antecedent and consequent of the causal conditional connected to each other by a counterfactual link. This modified analysis of causality was initiated by Reichenbach (1947, 1954) and has become more or less standard in discussions of causality within analytic philosophy and, after deeper modifications, in computational science (see Pearl 2000).

place, whether it is at all repeatable, whether it actually takes place only once, on all that the law is silent.

A special and important kind of the abuse of definitions in philosophy is the *circular* definition, in which the concept to be defined is presupposed in the definition. I have given some examples above in a different context. The fallacy of circularity in definitions comes about quite easily because people are not conscious that, in advance of the definition, they already possess the concept to be defined, precisely because it is not a concept created by definition but one originally given; and so it happens that, without noticing it, it is presupposed in the definition, precisely because people are not conscious of it. A case in point is that doctrine of which I spoke at the beginning of these lectures, the definition of TRUTH given by conventionalists such as Poincaré (one that he uses to presuppose the truth of a law of nature). Another case in point is the hedonistic definition of GOOD, which underlies the ethics of both Bentham and Mill (one that uses PLEASURE to define GOOD). To what extent these are circular definitions and examples of the abuse of definition in philosophy shall be the topic of our next lecture.

References

Bencivenga, Ermanno. 2000. *Hegel's dialectical logic*. New York: Oxford University Press.
Bergson, Henri. 1903. Introduction à la métaphysique. *Revue de métaphysique et de morale* 11: 1–36. [English translation: *Introduction to metaphysics*, New York, Putnam, 1912].
Berkeley, George. 1710. *A treatise concerning the principles of human knowledge*, Part I. Dublin: Aaron Rhames for Jeremy Pepyat.
Grelling, Kurt, and Leonard Nelson. 1908. Bemerkungen zu den Paradoxien von Russell und Burali-Forti [Remarks on the paradoxes of Russell and Burali-Forti]. *Abhandlungen der Fries'schen Schule (N.F.)* 2: 301–334. [Reprinted in Nelson (1971–1977), vol. III, pp. 95–129].
Locke, John. 1690. *An essay concerning humane understanding*. London: The Basset.
Nelson, Leonard. 1908. Über das sogenannte Erkenntnisproblem [On the so-called problem of knowledge]. *Abhandlungen der Fries'schen Schule (N.F.)* 2(4): 413–818. [Reprinted in Nelson (1971–1977), vol. II, pp. 59–393].
Nelson, Leonard. 1971–1977. *Gesammelte Schriften*, 9 vols. Edited by Paul Bernays, Willy Eichler, Arnold Gysin, Gustav Heckmann, Grete Henry-Hermann, Fritz von Hippel, Stephan Körner, Werner Kroebel, and Gerhard Weisser. Hamburg: Felix Meiner.
Pearl, Judea. 2000. *Causality: Models, reasoning, and inference*. New York: Oxford University Press.
Reichenbach, Hans. 1947. *Elements of formal logic*. New York: Macmillan.
Reichenbach, Hans. 1954. *Nomological statements and admissible operations*. Amsterdam: North-Holland.
Rosch, Eleanor H. 1973. Natural categories. *Cognitive Psychology* 4(3): 328–350.
Russell, Bertrand. 1903. *The principles of mathematics*, vol. I. Cambridge: University Press. [No second volume was ever published].
Wittgenstein, Ludwig. 1953. *Philosophische untersuchungen. Philosophical Investigations*. German text and English translation by G.E.M. Anscombe. Oxford: Blackwell.

Lecture XIII

Abstract The old (logical, deductive, so-called 'geometric') method is no use in philosophy; but the new one (regressive, analytic) 'axiomatic' method is. For it is none other than the 'critical' method invented by Kant. Mathematicians such as Hilbert have developed the axiomatic method to such perfection that we should learn from them to come to solid results in philosophy. By following the old mathematical ('dogmatic') method—a method which is perfectly appropriate to mathematics due to the nature of their objects of study—philosophers will fall again and again into the fallacy of concept-swapping.

In my last lecture I used an example from arithmetic to clarify the regressive method by which we uncover principles. That caused some confusion in the audience, as I also said that the regressive method befits philosophy in contradistinction to mathematics. But my example was deliberately taken from mathematics because I wanted to stay in the field of discussion. The confusion can be easily dissolved. The regressive method (which Kant called the *critical* method) is opposed to the so-called geometric method (Kant called it more appropriately the *dogmatic* method). The critical or regressive method, indispensable in philosophy, can nonetheless be used in mathematics as well. Its use in mathematics is in fact quite advantageous, albeit not of itself conducive to making new mathematical discoveries.[1] In philosophy we depend entirely on this method if we want to attain certainty and truth, not so in mathematics. Its use in mathematics is rather limited to developing this science in a rigorous, systematic form, i.e. to satisfying the logical ideal of systematic rigour in mathematics.

These two problems, the problem of the truth of a proposition and the problem of its provability, ought to be sharply distinguished. The difference between them should actually be clear from the previous discussion. For we have already seen that every proof presupposes unproven and even unprovable propositions. If these

---

[1]This opinion was widely held in Nelson's time, even though, as a matter of historical fact, it runs contrary to the whole mathematical tradition from the Ancient Greeks to Descartes and Leibniz. In our own times it was the great merit of György Pólya to bring back the old conception of analysis as an *ars inveniendi* under the name of 'heuristics' and 'plausible reasoning' (Pólya 1945, 1954, 1962).

unproven and even unprovable propositions are not true, then neither can the proven ones be true. Hence, if proving a proposition can never be sufficient to decide whether it is true, then neither can its unprovability be a criterion for its not being true. We cannot directly decide whether a proposition is true by proving it. But for some propositions we have no other way of convincing ourselves that they are true except by reducing them by means of proof to one or several antecedent propositions whose truth is directly known to us. Now, the clarity and self-evidence characteristic of geometry has the consequence that we can make good progress in attaining truth without having to prove anything. For proving mathematical truths is an exclusive feature of the history of European culture, not at all a general human preoccupation. The Indians, for example, display a geometry in which propositions are sequentially arranged without any proof, and they have no problem with that.[2]

The interest in proofs is thus an additional logical interest that has a distinct origin and should be sharply distinguished from an interest in truth as such. Mixing up these two demands has caused useless disputes—and a misplaced scepticism about mathematical axioms. The sceptical reasoner recognises a trivial fact, viz that mathematics relies on unprovable presuppositions, and so he commits an obvious fallacy—arguing that one should therefore deny them all truth and epistemic content. We often find this fallacy in the teaching of mathematics. Quite a few teachers, trying to instill truthfulness in their students, introduce them to mathematics by the exclusive use of rigorous proofs—a pedagogical mistake, for what can be easily understood without proof becomes opaque when we go the roundabout way of proof.[3] It is indeed generally much easier to gain insight into a simple mathematical proposition (say that $2 \times 2 = 4$) without proof than it would ever be by proving it. The aforesaid fallacy has its mirror image in the idea that truth is only known through proof, from which it would foolishly be concluded that we should first teach things in accordance with a false method [i.e. without proof] so as to enhance understanding [before using proof as the correct method].[4]

---

[2]Historians of mathematics seem to agree that logical rigour was a Greek invention (see Vega Reñón 1990; Netz 1999) and that Eastern mathematics, though in many ways more advanced, had no interest in proof. For a survey of the Indian case see Joseph (2011), compare Raju (2007).

[3]Nelson might have in mind the preface of Hermann Grassmann's *Handbook of Arithmetic* (1861, Preface, p. V): "Few people will contest that already for the first scientific instruction in mathematics the most rigorous method possible is to be preferred over any other. In particular, every schoolteacher will prefer a logically valid proof over any fallacious or circular one. Moreover, he will think it ethically unacceptable deliberately to present a proof of the latter sort to his pupils so as to deceive them to some extent. Nonetheless, this objectionable kind of so-called proofs have so far been the rule in all handbooks of arithmetic when teaching the foundations and the systematic articulation of the subject."

[4]The reader will see that Nelson's argument concerns the whole of mathematics and not only geometry. In his own time there were concerted efforts to perfect the axiomatisation of geometry as well as to bring axiomatic order into those parts of mathematics (arithmetic, algebra, analysis, number theory) that were not traditionally presented in that way; see Chapter "Lecture XVII". What the role of intuition in arithmetic according to Kantian and post-Kantian philosophy is supposed to be has always been somewhat obscure, which explains why, when talking about intuition, Nelson prefers examples from geometry.

Lecture XIII

Let us leave the teaching of mathematics aside and return to our question concerning the relationship between the methods of philosophy and geometry. I already said that for the purposes of a rigorously scientific construction the critical method not only *can* be applied in mathematics but it *must* be. A rigorously scientific construction requires that whatever can be proved must actually be proved. That is, of course, the demand logic makes on science. From a logical point of view it is not enough that the propositions from which we start are true. If we really want to start from them without proof, then we ought to be certain that they cannot be proved. The logical problem is how to reduce a given science to the smallest number of axioms, so that every one of them is necessary and all together are sufficient to construct that science.

It is for the purpose of analysing proofs that the regressive method shows its utility and fruitfulness. We can only get closer to the ideal of systematic rigour by analysing proofs in order to find out their underlying presuppositions, some of which we are not even conscious of using. With respect of each presupposition unearthed we must distinguish which one is necessary and which one is dispensable for this or that proof. This procedure allows us actually to prove propositions which are not only immediately self-evident but also *prima facie* incapable of proof. An example of one such immediately self-evident proposition is this:

(1) A continuous curve that runs between a negative and a positive point takes at least once the value zero.

The proof of (1) is superfluous in that its truth is immediately self-evident. The proof is also relatively complicated though it can be given. Or take a proposition which we discussed before and believed to be an axiom:

(2) The straight line is the shortest distance between two points.

Proposition (2) was quite artfully proved by Hilbert in his *Foundations of Geometry* (1899). And so was a proposition that from Euclid on had been regarded as an unprovable axiom:

(3) All right angles are equal.

Those are examples of geometric propositions that can be proved, even though they need not be proved if all we are interested in is whether they are true.[5]

We have thus come to distinguish between true axioms, propositions that cannot be proved, and apparent axioms, propositions that we have managed to prove—by discovering new axioms. So the number of axioms has decreased in one sense, yet it

---

[5] The theorem 'The straight line is the shortest distance between two points' was actually proved shortly before the publication of *Foundations*, in Hilbert (1895). The theorem 'All right angles are equal (congruent)' is proved in Hilbert (1899) as a consequence of the axioms of congruence. As for the idea that it is not necessary to prove what is self-evident, today's consensus among mathematicians is in profound disagreement with Nelson. For Hilbert's take on intuition see Hilbert and Cohn-Vossen (1932); for more discussion about the relationship between intuition and axiomatics see Carson and Huber (2006).

has increased in another sense, for we have discovered that some proofs have presuppositions whose existence we were not aware of. It was only a relatively short time ago that certain axioms in geometry were discovered—but not because we had doubts about their truth. On the contrary, they are so self-evident that we had not consciously singled them out as presuppositions in their own right. This is the case for example with propositions about the BETWEEN relation, such as (4) or (5):

(4) Between two points lying on a straight line there is always a third point.
(5) Of three points on a straight line one and only one point always lies between the other two.

Indeed, the axiomatisation of arithmetic (the logical construction of this science by means of an explicit small number of axioms) was first carried out in the nineteenth century.[6] The axiomatic ideal reached by Euclid for geometry has thus been attained by arithmetic after many centuries.

What is now important is to show that the method just described—the method to which the nineteenth century mathematicians were led in order to fulfill the logical ideal of systematic rigour in constructing their science—is none other than the method Kant demanded of philosophy, the one he called *critical*. The whole relationship between the two sister disciplines of geometry and philosophy was changed in the nineteenth century because of this development in mathematics, so that we have now a completely different situation from the one Kant experienced. We can now rightly say that philosophers would do well to imitate mathematicians in order to attain the kind of scientific rigour which commands universal admiration in the case of mathematics. To achieve that kind of rigour philosophers need only to take the so-called axiomatic method—the inquiry into axioms—and apply it to philosophy.[7]

Philosophers should not imitate the dogmatic method, the only one known to mathematics in Kant's time—for Kant has forever shown this to be a blind alley—but the critical method that mathematicians have come to use in their axiomatic inquiries. This is the method that Kant has shown philosophy needs. That method has in the meantime been developed by mathematicians into an admirably elaborated instrument, and the best a philosopher can do is to borrow this instrument

---

[6]The axiomatisation of arithmetic can be said to start with Hermann Grassmann's *Handbook of Arithmetic* (1861), although the first axiomatic systems were given in the 1880s (Peirce 1881; Dedekind 1888; Peano 1889). Frege's attempt to reduce the arithmetic axioms to logical ones (Frege 1893) was shattered by the discovery of Russell's paradox (Russell 1903, Chap. X; see Frege 1903).

[7]Although there are precursors of this idea of applying the axiomatic method to philosophy (the most obvious examples being Descartes, Spinoza, Leibniz, Bolzano and Peirce), it is worthwhile emphasising that Nelson has in mind the *modern* axiomatic method as first developed by Hilbert (1899). In Part IV of his *Critique of Practical Reason* (1917) Nelson in fact presented a pioneering application of axiomatics to the realm of ethics (probably the main reason he dedicated that book to Hilbert). Other early examples regarding different realms of philosophy are Ernst Mally's *Basic Laws of Duty* (1926), Rudolf Carnap's *Logical Structure of the World* (1928), and most successfully Alfred Tarski's *Concept of Truth in Formalised Languages* (1935).

from mathematicians, let them teach him how to use it, and practise it in their own work. Philosophy would thereby develop a different, more definite form than it has at present. The particulars of this lecture are, and intend to be, nothing more than examples by which I would like to convince you of the truth of this proposition.

I would like to run through a few names of philosophical writers so you can see for yourselves that I am talking about actual fallacies [not straw men]. In my last lecture I mentioned 'fluid concepts', required by a certain school of thought to avoid the defect of rigidity of ordinary scientific concepts. The alleged defect is apparently responsible for science being unable to do justice to the flux of real phenomena—to the vivacity and mobility of things.

The most celebrated writer of this school is the French philosopher Bergson. I refer you to his *Introduction to Metaphysics* (1903), where he develops this view very clearly and in a particularly elegant literary manner. Bergson's ideas entangle him in a polemic against mathematical physics. He objects that mathematical physics is fundamentally incapable of doing justice to its appointed task, for it faces an unsolvable problem as it wants to grasp continuous movement by means of rigid concepts. Continuous movement—a flux, a flow, or however else we call it—is, according to Bergson, an immediate experience of intuition for us, and no physical concept can ever actually grasp what is thus given in intuition, for the concepts of physics are of course rigid. How could they possibly be applied to flux? Bergson argues that physics proceeds like modern cinema, taking a sequence of shots of moving bodies, yet cannot have access to continuous movement itself. For that purpose it would need the fluid concepts of Bergson's metaphysics.

Although analysing his argument in detail would take us too far afield, the underlying fallacy is easy enough to uncover. The problem Bergson considers unsolvable was solved by physics a long time ago. It is essentially the problem discovered by Zeno in his paradoxes of movement—the paradoxes of the flying arrow and of Achilles and the turtle. Differential calculus gave physics the solution. An uncomplicated example is as follows: If we have a tangent *PQ* on a curve at point *P*, then the secant *PR* cutting the curve at point *R* changes continuously into *PQ* as it revolves around *P* (see Fig. 1).

This is a fact of intuition. Of course, one cannot say that the *concept* of SECANT changes into the *concept* of TANGENT. These concepts are indeed rigid, they cannot change into each other. In every position the revolving straight line is either still a secant or already a tangent. Bergson's objection to mathematical physics is nevertheless unjustified, for the TANGENT is defined as the mathematical LIMIT OF THE SECANT, and one can quite easily carry out the mathematical PASSAGE TO THE LIMIT from the secant to the tangent by using these concepts of differential calculus. This is therefore not an unsolvable problem at all, and the only fallacy would be if one said that the methods of differential calculus make intuition unnecessary. What we do here is to create concepts for what is given in intuition, so that every possible problem can be solved with the aid of these concepts. This is all that physics does, and intuition remains alongside the conceptual representation. But it remains as intuition, not as a cognitive operation that would use fluid concepts instead of the rigid ones of physics to get closer to its objects. For intuition is a non-conceptual

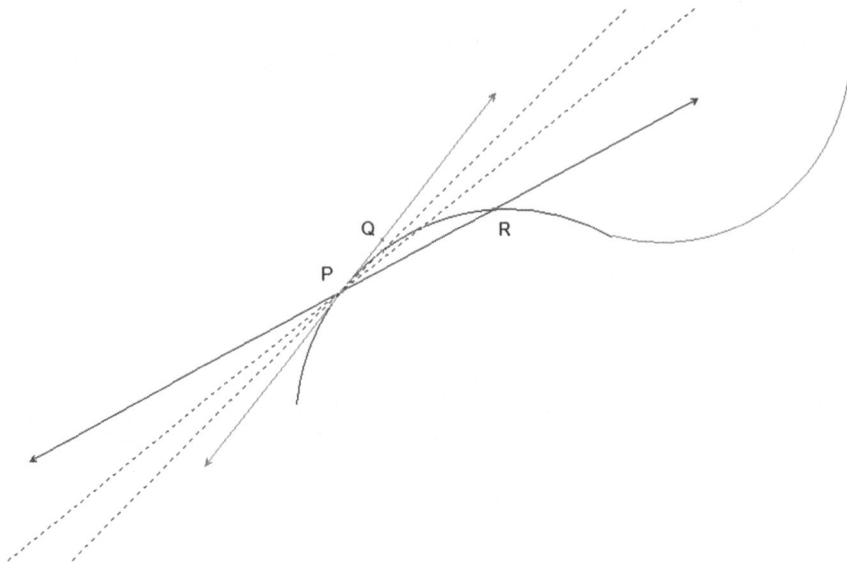

Fig. 1 A mathematical answer to Zeno's paradoxes of movement

cognitive operation; it does not operate with concepts, neither with rigid ones nor with fluid ones, which Bergson claims to have invented to this end.[8]

In my last lecture I also mentioned the doctrine of the 'universal self' or of 'consciousness in general' as examples of the still extant medieval idea of general objects. Amongst contemporary philosophers this idea is most poignantly present in Rickert, and especially in his book *The Object of Knowledge.* If you are interested in this topic, I would recommend the first edition, first because it is considerably shorter, thus saving time and energy, and secondly because its exposition of this idea is clear and consistent, while the later editions are burdened with an eclecticism that blurs the clarity and consistency of the first edition.[9] On the other hand, the later editions have the added benefit of presenting examples of another fallacy I mentioned, the concept of INDIVIDUAL CAUSALITY. This is one of those historical categories Rickert thinks it necessary to introduce besides the categories of natural science.

A similar view is to be found in Simmel's writings, in which he takes it over to the field of ethics. Simmel does not speak of 'individual causality' but of

---

[8] See Bergson (1903). It might be interesting to recall that some philosophers have claimed that geometry can be done without any appeal to intuition (and the diagrams that illustrate it). For a good treatment of this question see Greaves (2001) and for a wonderful discussion of the role of intuition in mathematics see Hadamard (1945).

[9] One can see what Nelson means by comparing the size of the three editions: 91 pages for the first edition (1892), 244 pages for the second (1904), and 456 pages for the third (1915).

'individual law', i.e. ethical law.[10] The 'individual law' is a law for the individual, for the individual in contrast to everyone else. The uniqueness and individuality of the individual, as Simmel contends, means that he is subject to a law that is valid for him alone, a law that prescribes his purpose in life, a purpose that he cannot share with any other individual.

Now, if what is meant by this is that in consideration of the individual uniqueness of this or that person he has certain tasks that others do not have, we have no objection to the idea; but it does not follow that there is an individual law, only that the particular tasks of this or that person emerge from applying the general law to a particular person. The application of the law generates a difference according to the particularities or even uniqueness of the case. The law itself remains one and the same, and the fallacy arises from failing to recognise the hypothetical form in which the general law alone can be expressed. The law demands that *if* a person has a particular feature F_1, then he should fulfill such and such tasks, whereas *if* a person has a different particular feature F_2, then he should fulfill different tasks. The laws of nature operate in the same way. Thus the law of gravity prescribes that two different bodies experience different accelerations depending, for instance, on the different distances between each one of them and a third body. But Simmel actually believes in the individualisation of the law for each individual person. The individual carries his own law in himself, which does not apply to any other person, not even to someone who has the same particular feature, and this is where the contradiction lies. The concept of LAW implies a necessary and distinct prescription, which we throw overboard the moment we demand something different of a person who shares a feature with another person. If *A* and *B* have tasks to fulfill, then from the fact that they have equal particular features it would necessarily follow that they both have to fulfill the same task. They could only differ numerically. The one person would of course not be the same person as the other. But that there are two of them would never be sufficient ground to distinguish between the ethical demands that would be placed on one or the other. If we place any kind of demand on a person, then it follows analytically that the same demands apply for every person who has the same features and is in the same situation. The concept of INDIVIDUAL LAW contains a contradiction.

In Chapter "Lecture III" another useful example came up—Poincaré's view on physical truths and the conflict he became involved in with his follower Le Roy. Let us remind ourselves what is it that we want to illustrate with our examples. We are looking for examples of the typical fallacy which stems (as I have shown) from the careless use of definitions in philosophy—careless in that before any definition is given we already have the concepts to be defined, so that an arbitrary definition always puts us into the precarious situation of confusing the concept introduced by the definition with the concept that we had before, and thus committing all the fallacies that follow from this confusion, as set forth in Chapter "Lecture XII". One such example is Poincaré's definition of TRUTH. This is a concept that we possess

---

[10]See Simmel (1905, Chap. 2, 1918, Chap. IV).

before ever trying to define it, not one that we first become acquainted with by means of a definition.

This kind of definition is not only to be found in Poincaré but is common to a whole school of thought that has become quite widespread in our own time. The philosophers of that school are called, or they call themselves, 'pragmatists'. The name is quite appropriate, for what these philosophers intend is to reduce the concept of TRUTH to a practical concept—the concept of CONVENIENCE. A physical statement about a law of nature according to Poincaré means that we find it convenient to state it. I have already shown that this definition involves a logical circle. It is logically circular, insofar as the concept of truth is already presupposed in the definition. The wording which makes this perfectly clear is as in (6):

(6) The truth of a judgment, of holding something to be true, consists in its being convenient.

We immediately see that the concept of TRUTH is already presupposed in definition (6), for we are thereby holding something to be true, viz that of which our holding it to be true is said to be convenient, and we cannot make such a statement unless we presuppose the concept of TRUTH in making it. This presupposition, this concept, is to be found in every 'is' or 'there is' (or in any other wording we might want to use), for we cannot avoid using some such expression, independently of the content of our judgments. If we want to keep the meaning of such expressions, if we do not want to dabble in meaningless expressions that nobody will consider convenient, then their meaning can only be what the word 'truth' itself expresses. Let us make the statement more precise:

(7) It is true that the truth of a physical statement only lies in its convenience.

No matter what judgment we utter, it will always analytically contain the statement 'It is true that…', or 'It is the case that…'—whatever we are stating. Hence it follows that this definition [or any other definition of TRUTH for that matter] already presupposes a different concept of TRUTH than that of CONVENIENCE, at least if it has any kind of meaning and we are not merely dabbling in empty words. A pragmatist can naturally escape the objection by saying: 'I want to use the word *truth* in the sense of CONVENIENCE.' That is of course allowed. But in that case he should not forget that by so changing the meaning of words they have not eliminated the concept that was previously associated with the word 'truth', a concept that other people still associate with the word, indeed a concept that he himself still uses in the very act of redefining it, for in doing so he is of course expressing a judgment.[11] The pragmatist is informing us that he wishes to use the word 'truth' in the sense of

---

[11]Nelson was perfectly aware that a nominal definition that stipulates how a given term shall be understood does not have a truth value. It is just a convention. But the whole point here is that the pragmatist definition is not presented as a mere convention but as somehow capturing the 'essence' of truth. So Nelson's argument is that such a claim *is* false, because it contradicts itself. The point is taken up at the beginning of Chapter "Lecture XVI" when the question is raised of how the concept of STATE should be defined.

the word 'convenience', and that is supposed to be true, not just true-in-the-sense-of-convenient. The contradiction that follows from such a circular definition is thus plain to see.[12]

References

Bergson, Henri. 1903. Introduction à la métaphysique. *Revue de métaphysique et de morale* 11: 1–36. [English translation: *Introduction to metaphysics*. New York: Putnam (1912)].
Carnap, Rudolf. 1928. *Der logische Aufbau der Welt*. Leipzig: Felix Meiner. [English translation: *The logical structure of the world: Pseudoproblems in Philosophy*. Berkeley: University of California Press (1967)].
Carson, Emily, and Renate Huber, eds. 2006. *Intuition and the axiomatic method*. Dordrecht: Springer.
Dedekind, Richard. 1888. *Was sind und was sollen die Zahlen?* Braunschweig: Vieweg. [English translation: The nature and meaning of numbers, in *Essays in the theory of numbers*, pp. 31–115. Chicago: Open Court (1901)].
Frege, Gottlob. 1893. *Grundgesezte der Arithmetik, begriffsschriftlich abgeleitet*, vol. I. Jena: Hermann Pohle. [English translation: *Basic laws of arithmetic*. Oxford: University Press, 2013].
Frege, Gottlob. 1903. *Grundgesezte der Arithmetik, begriffsschriftlich abgeleitet*, vol. II. Jena: Hermann Pohle. [English translation: *Basic laws of arithmetic*. Oxford: University Press (2013)].
Grassmann, Hermann. 1861. *Lehrbuch der Arithmetik für höhere Lehranstalten* [*Handbook of arithmetic for institutions of higher education*]. Berlin: Enslin.
Greaves, Mark. 2001. *The philosophical status of diagrams*. Stanford: Center for the Study of Language and Information.
Hadamard, Jacques. 1945. *An essay on the psychology of invention in the mathematical field*. Princeton: University Press.
Hilbert, David. 1895. Ueber die gerade Linie als kürzeste Verbindung zweier Punkte [On the straight line as the shortest connection between two points]. *Mathematische Annalen* 46: 91–96.
Hilbert, David. 1899. *Grundlagen der Geometrie*. Leipzig: Teubner. [English translation: *Foundations of geometry*. Chicago: Open Court (1971)].
Hilbert, David, and Stephan Cohn-Vossen. 1932. *Anschauliche Geometrie*. Berlin: Julius Springer. [English translation: *Geometry and the imagination*. New York: Chelsea Publishing Co. (1952)].
Joseph, George G. 2011. *The crest of the peacock: Non-European roots of mathematics*, 3rd ed. Princeton: University Press.
Mally, Ernst. 1926. *Grundgesetze des Sollens: Elemente der Logik des Willens* [*Basic laws of duty: elements of the logic of the will*]. Graz: Lauschner & Lubensky [Reprinted in *Logische Schriften*, eds. Karl Wolf and Paul Weingartner, 229–324. Dordrecht: Reidel (1971)].
Nelson, Leonard. 1908. Über das sogenannte Erkenntnisproblem [On the so-called problem of knowledge]. *Abhandlungen der Fries'schen Schule* (N.F.) 2(4): 413–818 [Reprinted in Nelson (1971–1977), vol. II, 59–393].
Nelson, Leonard. 1917. *Kritik der praktischen Vernunft* [*Critique of practical reason*]. Göttingen: Vandenhoeck & Ruprecht [Reprinted in Nelson (1971–1977), vol. IV].

---

[12]In Nelson (1908) the pragmatist definition had already been criticised, practically at the same time as Russell published his own in many respects similar critique (Russell 1908).

Nelson, Leonard. 1971–1977. *Gesammelte Schriften*, 9 vols. eds. Paul Bernays, Willy Eichler, Arnold Gysin, Gustav Heckmann, Grete Henry-Hermann, Fritz von Hippel, Stephan Körner, Werner Kroebel, and Gerhard Weisser. Hamburg: Felix Meiner.

Netz, Reviel. 1999. *The shaping of deduction in greek mathematics: A study in cognitive history*. Cambridge: University Press.

Peano, Giuseppe. 1889. *Arithmetices principia nova methodo exposita*. Torino: Bocca. [English translation: The principles of arithmetic, Presented by a New Method, in Jean van Heijenoort, *From Frege to Gödel: A Source Book in Mathematical Logic, 1879–1931*, 85–97. Cambridge (MA): Harvard University Press (1967)].

Peirce, Charles S. 1881. On the logic of number. *American Journal of Mathematics* 4: 1–4.

Pólya, György. 1945. *How to solve it: a new aspect of mathematical method*. Princeton: University Press.

Pólya, György. 1954. *Mathematics and plausible reasoning*, 2 vols. Princeton: University Press.

Pólya, György. 1962. *Mathematical discovery: on understanding, learning, and teaching problem solving*, 2 vols. New York: John Wiley.

Raju, C.K. 2007. *Cultural foundations of mathematics: The nature of mathematical proof and the transmission of the calculus from India to Europe in the sixteenth century C.E.* New Delhi: Pearson Longman.

Rickert, Heinrich. 1892, 1904, 1915. *Der Gegenstand der Erkenntnis* [*The object of knowledge*]. Three increasingly larger editions. Tübingen: Mohr.

Russell, Bertrand. 1903. *The principles of mathematics*, vol. I. Cambridge: University Press [No second volume was ever published].

Russell, Bertrand. 1908. Transatlantic 'Truth'. *The Albany Review* 2: 393–410.

Simmel, Georg. 1905. *Die Probleme der Geschichtsphilosophie: Eine erkenntnistheoretische Studie* [*The problems of the history of philosophy: an epistemological study*], 2nd ed. Munich and Leipzig: Duncker & Humblot.

Simmel, Georg. 1918. *Lebensanschauung: Vier metaphysische Kapitel*. Munich and Leipzig: Duncker & Humblot. [English translation: *The view of life: four metaphysical essays with journal aphorisms*. Chicago: The University of Chicago Press (2010)].

Tarski, Alfred. 1935. Der Wahrheitsbegriff in den formalisierten Sprachen. *Studia Philosophica: Commentarii Societatis Philosophicae Polonorum* 1: 261–405. [This is a German translation, by Leopold Blaustein, of the original Polish text published in 1933. English translation from the German: 'The concept of truth in formalized languages', in *Logic, semantics, metamathematics: papers from 1923–1938* (152–278), Oxford, Clarendon Press (1956)].

Vega Reñón, Luis. 1990. *La trama de la demostración* [*The warp and woof of proof*]. Madrid: Alianza Editorial.

Lecture XIV

Abstract The concept-swapping fallacy produces circular definitions and infinite regresses. It is found in theoretical disciplines, as shows the case of Poincaré and Le Roy in the philosophy of mathematics. But it is also found in practical ones. Thus in ethics we find it in the repeated attempts to define the good, by authors as different as Bentham and Mill, on the one hand, or Brentano on the other. And in the philosophy of law we find it in the convoluted ways in which legal positivists attempt to define what is lawful. In all these and many other cases, what is at work is the replacement of a real synthetic judgment by an apparent analytic one.

At the end of my last lecture I was critising Poincaré's (1902) definition of TRUTH as an example of the typical fallacy we have been discussing for quite a while—a fallacy that is the main general form to which the particular fallacies that play a role in philosophy can be reduced. More on that point presently. The fallacy consists in arbitrarily defining a concept that we are already acquainted with. When we do that, it often happens that the definition is circular in that the original concept is hidden in the *definiens*.

The fallacy can also be shown to be a case of infinite regress, as I demonstrated when this example first came up (see Chapter "Lecture III"). If we replace the concept to be defined [TRUE] by its *definiens* [CONVENIENT] in the propositions in which the former occurs, we would be expressing the meaning captured by the definition. What happens, however, is that such replacement has to be repeated, because the original concept sneaks in again, so that the defining procedure ends in an infinite regress and we never capture the meaning intended. Those sentences [in which the concept of TRUTH occurs] would only become meaningful if the regress would come to an end somewhere. Mathematics offers us well-known and meaningful infinite processes, but never concerning the contents of a definition. Otherwise we would never associate any meaning with the words [or symbols] we use. Meaning is lost in an infinite regress, for if every link in the chain first gets its meaning from the one that follows, and if that chain is infinite, then the whole chain of words lacks any meaning.

It is interesting and instructive to return to the contrast between Poincaré and his follower Le Roy that this issue brought to light (see Chapter "Lecture III"). Le Roy

(1899, 1900) makes use of Poincaré's physical pragmatism, as you will remember, in order to demonstrate that the attempt by physicists to arrive at true propositions about nature is in vain. Poincaré says that the meaning of physical truths is convention, yet this is not at all Le Roy's position. Le Roy assumes only the negative part of Poincaré's thesis, viz that physics can only yield conventions about nature— posited arbitrarily by us. Since Le Roy, however, maintains the original concept of TRUTH, he uses it to conclude that physics cannot yield any truth at all.[1] He thus presupposes that there is truth in the original sense yet physics is not able to arrive at it. Truth is rather reached through the authority of the church and its tradition. Hence only physical propositions are untrue, whereas those of dogmatic theology are true. That is why I say that Le Roy, in contrast to Poincaré, firmly maintains the original concept of TRUTH.

Notice, however, that Le Roy's argument can also be shown to contain a contradiction. His view is not only false but inconsistent, and if Poincaré had noticed it, he would have been able to ward Le Roy off. Both Poincaré and Le Roy overlook that a contradiction is lurking in the very question,

(1) What is the meaning of the statement that physical propositions are nothing but convenient assumptions?

If one posits a natural law in Poincaré's sense, then no truth is asserted about nature as it is and surrounds us, but all we are allowed to say is (2):

(2) It is convenient to introduce this or that assumption—this or that natural law.

Let us now clarify what 'convenient' means. What we call 'convenient' is a means with regard to an end we try to achieve. But what is that—a means relative to an end? An end is just an effect we try to achieve, and a means is just an event which is the sufficient cause for the effect to occur. A means is convenient if it helps bring about an effect we desire. Hence asserting that an event, e.g. an assumption, is convenient implies asserting that it is a cause, that it helps bring about an effect we desire. And so we have asserted a connection of one event with another, we have asserted that the one event is the cause of the other event and the latter the effect of the former. In other words, we have presupposed a natural law on which the means-end connection is based—a connection that gives us the right to talk about means and ends, and to talk about convenience. Every statement about an event being convenient implies asserting a natural law and without this presupposition it loses all its meaning. Therefore, the doctrine that our beliefs about the existence of natural laws are not true in the original sense of the word, but that they are just a matter of convenience, is inconsistent. It presupposes what it rejects.

Let us consider another example of the abuse of definitions in philosophy. For this we will also draw on a fallacy that we pointed out earlier, viz the hedonistic

---

[1] The most relevant parts of Le Roy's argument are probably his 'criticism of scientific facts' (1899, pp. 514–518), his 'criticism of the laws of nature' (*ibid* 518–526) and his 'criticism of physical theories' (*ibid* 526–534). It is also remarkable that the author, when talking about the experimental method, says that truth is no more than 'the inner joy' of the scientist (*ibid* 512).

definition of GOOD, which I mentioned in connection with the pragmatistic fallacy. GOOD is as much a concept that is given before any definition as is TRUE. So the argument about the concept of TRUTH in physics is as valid as the corresponding argument about the concept of GOOD in ethics. I referred to Bentham (1789), who defines GOOD as PLEASURE [see Chapter "Lecture III"]. Similar definitions are quite frequent in ethics. GOOD has for example been defined as WHAT WE DESIRE, WHAT WE STRIVE FOR, or just WHAT WE LIKE. In all these definitions of GOOD, the concept to be defined, the concept of GOOD—or what is actually the same thing, the concept of VALUE—is covertly presupposed. The word 'value' makes the circularity even clearer. Let us remind ourselves that all pleasure, all liking, all wishing, striving and desiring is just a specific form of the act of valuing. To take pleasure in something means to carry out a specific act by which we value the object in question, and the same is true of desire. And so whoever defines VALUE in one of the indicated ways is just defining it as 'what we consider or judge valuable'. You can see that in such a definition the concept of VALUE is presupposed.

The point here is not whether one assumes or rejects objective values, i.e. values which exist independently of our valuations. To be sure, the definitions we are discussing lead to ethical subjectivism, for they shift value to the subjective fact of valuation and make it identical with it. These definitions know no value independent of valuation. The consequence of this view is the denial of any kind of objective value—to say that something is valuable would be a psychological statement, a statement about certain conscious acts of valuation. We cannot speak of 'valuation' without presupposing the concept of value—even if the value does not exist objectively but is only imagined, even if the concept of VALUE lacks reference, we nonetheless presuppose it as a concept and use it covertly in the definition.

We are in all such cases deluded by words. This becomes clearer when instead of speaking of 'pleasure', 'desiring', 'liking', and so on, we just speak of 'valuation' and make thereby explicit the common element to all these psychological acts. It is self-evident that from such a subjectivistic perspective we cannot arrive at any kind of ethical norm, for such a norm would be a criterion for the distinction between valuations that are correct or incorrect, founded or unfounded. If VALUE is just defined as VALUATION, then it does not make sense to speak of correct and incorrect values, there is no objective standard. Anyone who believes in such a standard, anyone who wants to distinguish between what is actually good and what we just *think* is good, must constrain the theory we are discussing, and indeed this is exactly what Bentham's successor, Mill (1861), did. Any ethics requires a norm that allows for valuations that are correct [as opposed to incorrect]. Theory has to satisfy this need and Mill provides for it by distinguishing between higher and lower pleasures, between higher and lower satisfactions, true and false or imaginary happiness.[2]

Mill's modification of Bentham's theory makes the circularity all the more obvious, for it introduces a valuation of pleasure according to an independent standard, i.e. according to a concept of GOOD that comes from somewhere else.

---

[2] See Mill (1861, Chap. II).

What is a *higher* pleasure or a *true* pleasure if not being pleased by something good as opposed to being pleased by something bad or less valuable? Only that PLEASURE is a criterion of the GOOD which consists in BEING PLEASED BY THE GOOD. And how on earth are we to distinguish between being pleased by something good and being pleased by something bad without presupposing a concept of GOOD that is different from PLEASURE? Without any such presupposition we get an infinite regress in our definition. Being pleased by something good would be taking pleasure in the object of true pleasure, i.e. taking pleasure in what is good. And so our definition has incorporated the concept of GOOD which it was supposed to define. So we must again replace that concept with the *definiens*—and so on in an infinite regress which makes our definition meaningless.

In the history of modern ethics we find a sublime form of this fallacy in Franz Brentano's famous treatise *On the Origin of Moral Knowledge* (1889). He endeavours here to define the concept of GOOD so as to overcome ethical subjectivism and hedonism. Brentano defines it as in (3)[3]:

(3) We call something GOOD if LOVING IT IS RIGHT.

The word 'love' in (3) is rather infelicitous, for what Brentano means is liking or valuing. So we correct as in (4):

(4) We call something GOOD if VALUING IT IS RIGHT.

When is it right to value something? Only when it is good to value it [for 'right' and 'good' mean the same in this context]. We are moving in a circle again. Brentano could have defined 'good' just as easily and indeed more clearly by saying,

(5) We call something GOOD if IT IS AN OBJECT WE LIKE AND WHAT WE LIKE IS THE GOOD.

For that is in fact the meaning of the words if it is right to love the said object. If we substitute those words we arrive at (5). It is stunning that there is nowadays an entire school of philosophers, mostly 'phenomenologists', as they are called, who see in this discovery of Brentano's a far-reaching reform of ethics. Facts of this kind show that these are not mere sophisms. There must be a rather different and much deeper explanation of why otherwise astute thinkers become entangled in such crude fallacies. A particularly good illustration of the underlying problem is Brentano's mistaken belief that RIGHT LOVE is self-evident. This would imply that ethical knowledge is based on self-evident axioms of the sort we find in geometry. This is a false presupposition. The best way to refute such an assumption is to point to such crass errors about the most elementary concepts of a subject—ethics—that is alleged to be self-evident. Brentano's example shows plainly that the method of introducing concepts through definition is misleading and dangerous.

---

[3]See Brentano (1889, §23).

Let us take another example from a practical discipline.[4] The main concept in the study of law is—LAW.[5] Even today students of law argue over the correct definition of the concept of LAW. There are several definitions which have met with widespread approval. I would like to address two of them here. The first one defines the concept of LAW as ENFORCEABILITY OF A RULE OF CONDUCT.[6] This definition contains an ambiguity. Enforceability is the possibility of enforcing the rule in question. This possibility can be understood in a physical or in a legal sense. Hardly anyone will seriously consider defining LAW as PHYSICAL ENFORCEABILITY. This would imply that might is right, thus boliling it all down to the rule of the strong. If enforceability of a rule of conduct is not to be understood in a physical sense, then it has to be understood in a legal sense. Enforceability then is just the legal right to enforce a rule of conduct. Somebody exercises his legal right to enforce a rule of conduct only if he does not do so unlawfully.[7] So we must incorporate the lawfulness restriction into the definition, for not every enforcement of a rule of conduct is lawful. We get again a circular definition—we define LAW as

THE ATTRIBUTE A RULE OF CONDUCT POSSESSES WHEN IT IS LAWFULLY ENFORCED,

or as

THE LEGAL RIGHT TO ENFORCE A RULE OF CONDUCT.

The concept of LAW [or LEGAL RIGHT] that we would have to presuppose is different from the one defined, if we do not want to become entangled in an infinite regress. Otherwise, in defining LEGAL RIGHT TO ENFORCE, we would have to insert the *definiens*—with the consequence that LEGAL RIGHT TO ENFORCE would be defined as

THE POSSIBILITY OF ENFORCING THE ENFORCEMENT,

and such enforcement of the enforcement would have to be enforced and so on ad infinitum. Thus, anyone who wishes to avoid the infinite regress and a circular

---

[4]The German phrase 'practical discipline' (*praktische Wissenschaft*), at least as used in Kantian philosophy, means more or less the same as 'normative discipline'. Thus both ethics and the philosophy of law would be 'practical disciplines' in this sense.

[5]The following argument relies to a considerable extent on the double meaning of the German word *Recht*, which means both 'law' and 'right'.

[6]Legal positivists, however, have notoriously made statements much to that effect, e.g. Jellinek (1914, p. 333). A detailed discussion of Jellinek can be found in Nelson (1917, Chap. 1). Nonetheless, it is only fair to say that Jellinek (1914, p. 334, n. 2) admits that enforceability has to meet serious objections.

[7]The German text is somewhat opaque at this point. It seems possible, if a bit stretched, to translate it as follows, 'Enforcing a rule of conduct makes that rule a law only if the enforcement is lawful.' However, the translation given above, even if freer, reflects Nelson's argument better.

definition, will have to presuppose another concept of LAW [and LEGAL RIGHT], thus contradicting the given definition.[8]

A second and more modern definition of the concept of LAW, which most lawyers come up with after having recognised that the above definition is inadmissible, amounts to defining the LAWFUL character of a rule of conduct as its BEING RECOGNISED—especially by those subject to it.[9] The question then arises: as what is the said rule recognised? Obviously as LAW. Yet we are then once more caught in the same logical circle as before. To be sure, the rule in question has to be recognised as lawful and not simply as enforced if recognition should make it lawful. If we substitute the *definiens* we arrive at definition (6):

(6) A rule of conduct is lawful when it is recognised as having its lawful character recognised.

We find ourselves caught in the same circle, since the defined concept is contained in the definition. We find ourselves in an infinite series:

(7) Law is what is recognised as something which we recognise as being recognised, *and so on*.

In this way we never reach a recognition that makes the rule lawful.

The same fallacy can be seen even more clearly in another very popular formulation. It defines [the concept of] LAW as follows:[10]

AGREEING WITH THE CONVICTION OF LAWFULNESS.

In other words, the lawful character of a rule depends on the people or the majority (or whoever one thinks should be crucial) being convinced of its lawfulness. The very expression 'being convinced of its lawfulness' makes the circle perfectly visible. It is obvious that the lawfulness conviction is either right or wrong. If it is right, then this can only mean that the rule people are convinced of is right, but if it *is* right, then it would not cease to be right if people were by mistake not convinced of it. On the other hand, if the lawfulness conviction is mistaken, i.e. if what people are convinced of as being right is not right, then we have contradicted the definition that law has to agree with the lawfulness conviction. The

---

[8] Nelson did not take advantage of the logical method of definition by means of variables which Russell (1905) used to so much effect. If he had, his argument would have been less convoluted. He would have started with 'x is legal iff x is enforceable'. Then under exclusion of mere physical enforceability we would get 'x is legal iff x is legally enforceable'. The meaning of the adverb 'legally' would obviously become unreachable after an endless substitution of the *definiens*: 'x is legal iff x is ... enforceably enforceably enforceably enforceable'.

[9] The concept of recognition or similar ones were also popular with German legal positivists, e.g. Jellinek (1914, p. 371). Compare Nelson (1917, p. 49).

[10] The phrase 'conviction of lawfulness' translates German *Rechtsüberzeugung*, which refers to the *feeling* most citizens have about the *legitimacy* of laws, decrees or government policies. The somewhat inelegant translation chosen helps make Nelson's argument clear. See Jellinek (1914, pp. 333–334, 355) and Nelson (1917, p. 16).

only astonishing thing is that such a contradictory definition could have met with such widespread approval.

Through just a few examples you have come to recognise a type of fallacy that you will now easily discover for yourselves in other cases. They may be enough to illustrate the abuse of definitions in philosophy, an abuse that comes from overlooking the fact that we are dealing with concepts that we already have and not concepts that are first created by a definition. The abuse of definitions that has now become familiar to you explains an otherwise inscrutable paradox, it explains how it is possible for a kind of metaphysics to be erected whose method can lead nowhere, as it should have been clear from the start. How was it possible that dogmatic metaphysics could ever stage a play in which a scientific building seems to be erected, and in fact keeps staging the same play again and again? How can we thus deceive ourselves? How can we explain the illusion?

I shall now answer that question. The illusion arises because people take an analytic judgment, i.e. a judgment that is based on a definition and whose analytic character makes it indisputable, and unconsciously replace it with a synthetic judgment, and by so doing create the illusion of a fruitful scientific procedure.[11] Such an illusion is well known in a different field, for it is essentially what logicians call a *quaternio terminorum:* by replacing the concept that occurs in one premise by a different concept in the other premise we get an argument lacking a middle term. This is how an analytic judgment is transformed into a synthetic one. Although this fallacious inference is widely known in logic, it is not at all as dangerous there as it is in philosophy. Consider the following textbook example:

Herodes is a fox
Every fox has four legs

Herodes has four legs

If you just look at the *words* this argument has only three terms ['Herodes', 'fox', 'having four legs'], as it should, yet we plainly have here four *concepts*. People may in some cases easily overlook that two different concepts correspond to one single word ['fox' in this textbook example].

A somewhat less transparent example has been playing a role in the history of philosophy for quite some time and even today exerts its influence as far as the exact sciences—the problem of the possibility of action at a distance.[12] Even today many physicists see a logical impossibility in such an assumption and argue as follows: 'A body can only act locally—it has to be locally present where it exerts its

---

[11]Nelson touches here the difficult question of how a definition can actually *produce results*. This question was first adumbrated by Kant, who believed that this was only possible in the case of mathematics, where we were able to construct the mathematical object corresponding to the concept defined, because the definition gave us the relevant instructions. Frege (1884) was concerned with this question in all his work, starting with his *Principles of Arithmetic*. Nelson's point is that this works in mathematics but can and will never work in philosophy.

[12]This problem arose from Newton's theory of universal gravitation. For a historical overview of the conceptual difficulties of action at a distance, see Hesse (1962).

force.' But action at a distance is a force going from a body which acts on a distant body, i.e. not locally. This argument deceives quite a few physicists and thus blocks the path to a solution of the problem. It is clear that only experience can decide whether or not there is action at a distance in nature. But they decide the question in advance of experience on the basis of a logical pseudo-proof. The fallacy is the same as before when we argued that Herodes has four legs because he is a fox. 'A body can only act locally.' Fine, but this only means that it has to be locally present where it exerts its force, but it does not have to be locally present where the body acted upon is. These two concepts are mixed up because of the ambiguous expression 'locally present'.

References

Bentham, Jeremy. 1789. *An Introduction to the Principles of Morals and Legislation.* London: Payne.
Brentano, Franz. 1889. *Vom Ursprung sittlicher Erkenntnis.* Leipzig: Duncker & Humblot. [English translation: *The Origin of the Knowledge of Right and Wrong*, Westminster, Archibald Constable & Co., 1902].
Frege, Gottlob. 1884. *Die Grundlagen der Arithmetik: eine logisch-arithmetische Untersuchung über den Begriff der Zahl.* Wrocław: Wilhelm Koebner. [English translation by J.L. Austin: *The Foundations of Arithmetic: A Logico-Mathematical Enquiry into the Concept of Number*, Oxford, Blackwell, 1950. New translation by Dale Jacquette: *The foundations of arithmetic.* New York: Pearson, 2007].
Hesse, Mary B. 1962. *Forces and Fields: The Concept of Action at a Distance in the History of Physics.* London and New York: T. Nelson.
Jellinek, Georg. 1914. *Allgemeine Staatslehre* [General theory of the state], 3rd ed. Berlin: Oscar Häring. [First edition 1900, second edition 1905].
Le Roy, Édouard. 1899, 1900. Science et philosophie [Science and philosophy]. *Revue de métaphysique et de morale*, VII, 375–425, 503–562, 708–731; VIII, 37–72.
Mill, John Stuart. 1861. Utilitarianism. *Fraser's Magazine* 64: 391–406, 525–534, 658–673 [Reprinted many times].
Nelson, Leonard. 1917. *Die Rechtswissenschaft ohne Recht: kritische Betrachtungen über die Grundlagen des Staats- und Völkerrechts, insbesondere über die Lehre von der Souveranität* [*Law without justice: critical reflections on the foundations of public and international law, in particular on the theory of sovereignty*]. Leipzig: Veit & Comp. [Reprinted in Nelson (1971–1977), vol. IX, pp. 123–324].
Nelson, Leonard. 1971–1977. *Gesammelte Schriften*, 9 vols. Edited by Paul Bernays, Willy Eichler, Arnold Gysin, Gustav Heckmann, Grete Henry-Hermann, Fritz von Hippel, Stephan Körner, Werner Kroebel, and Gerhard Weisser. Hamburg: Felix Meiner.
Poincaré, Henri. 1902. *La science et l'hypothèse.* Paris: Flammarion. [English translation: *The Foundations of Science: Science and Hypothesis, The Value of Science, Science and Method*, New York, The Science Press, 1929, pp. 27–197].
Russell, Bertrand. 1905. On Denoting. *Mind* 14(56): 479–493.

Lecture XV

Abstract If concept-swapping is a fallacy consisting in arguing in favour of a synthetic statement disguised as an analytic one, then there is nothing more important for philosophical methodology than to have a firm and clear grasp of the analytic-synthetic distinction. A useful byproduct of this would be that we will be in a position to understand why serious scholars spend so much time arguing about definitions, thereby treating terminological questions as though they were substantive. Although this strange phenomenon can be observed in all areas of philosophy nowhere is there more at stake as when concept-swapping is taken to decide questions of law.

At the end of our last lecture we saw that dogmatic philosophy cannot decide whether the concepts that are its subject matter are empty or even contradictory. Dogmatic philosophers cannot decide this question because (a) they introduce those concepts via an arbitrary definition, and (b) they lack a cognitive source that would make it evident that the objects of those concepts do exist. In this lecture we will extend these considerations from the concepts of dogmatic philosophy to the analogous case of its propositions.

The question arises as to how on earth it is possible for dogmatic philosophy to make progress, to move from the spot, given that all it has to work with are the concepts it has itself arbitrarily created. We know that from mere concepts only analytic judgments can result, never synthetic ones. Dogmatic philosophy asserts that it can establish synthetic propositions, i.e. it claims to expand our knowledge. 'Metaphysics' is the name it gives to such an endeavour. How can we be deceived that expanding our knowledge by dogmatic methods is at all possible? That is the question we have asked and to which I now seek the answer.

We started an example-guided inquiry and know already in a general way that the illusion about a dogmatic extension of our knowledge can only arise by confusing analytic judgements with synthetic ones. This illusion, upon which the entire edifice of dogmatic metaphysics is erected, vanishes as soon as we distinguish between analytic and synthetic judgments. That is the ultimate importance of this distinction. Whoever is not sufficiently conscious of this distinction may easily form an analytic judgment on the basis of an arbitrary definition of a concept, substitute a synthetic judgment for the analytic one, and then deceive himself that he has thereby expanded his knowledge. In all such cases two concepts are swapped, there is a *quaternio terminorum*, and the conclusion will be a synthetic judgment, an unfounded and even false one.

In Chapters "Lecture IV" and "Lecture V" we discussed an example from the Schoolmen that I would like to resume here as it has a particular importance in the history of philosophy and is also very instructive at the present stage of our inquiry. I mean the proposition, 'Predicates do not oppose each other', which plays a role in the ontological proof of God's existence and is otherwise important for scholastic metaphysics. The concept of GOD was defined via the concept of the UNIVERSAL SET and the assumption that this concept of a UNIVERSAL SET has an object [has reference] presupposes the truth of the proposition, 'Predicates do not oppose each other.' If some predicates oppose each other, i.e. if they cannot occur together in one object, then there is no Universal Set—and yet a certain illusion beguiles us into believing that predicates do not oppose each other.

This illusion is based on mistaking the concept of OPPOSITION for the concept of CONTRADICTION. Earlier on we convinced ourselves that the proposition is true if thus formulated: 'Predicates do not *contradict* each other.' A contradiction only takes place between any predicate and its negation. If we confuse the two concepts of CONTRADICTION and OPPOSITION (the word 'opposition' meaning the mere incompatibility of two predicates in relation to one object, never mind whether or not it is a *logical* incompatibility, i.e. a contradiction proper), then the analytic judgment, 'Predicates do not contradict each other', is replaced by a synthetic judgment, 'Predicates do not oppose each other'; yet this synthetic judgment has not only been introduced without sufficient reason, but it is in fact false, as immediately seen by experience, witness the concept of a WINGED HORSE. This concept does not of course involve a contradiction, and yet the two predicates brought together in WINGED HORSE (viz HORSE and WINGED BEING) are opposed, they exclude one another in actual reality.

This example plays an important role in the philosophy of the Schools, but analogous examples are to be found in abundance in contemporary philosophy. I would take up a few. Only the other day a participant in this course very kindly handed me an issue of the journal *Triarticulation of the Social Organism*, doubtless in order to prepare me for an attack on my theories, planned for a Stuttgart conference in August of this year. I would like to pre-empt this attack by communicating my view on a paper published in that issue on the subject of Rudolf Steiner's *Philosophy of Freedom* (1894). Mr. Stein, the author of that paper tries to refute an argument against Steiner by one of his most recent critics, Mr. Kerler (1921).[1]

---

[1] The background for this story is a German esoteric movement called 'anthroposophy', whose leading spirit was Rudolf Steiner, a publicist and social and educational reformer very active at the beginning of the twentieth century. In the Anthroposophical Conference Nelson alludes to ('Cultural Perspectives of the Anthroposophic Movement', Stuttgart, from 28 August through 7 September 1921), a certain Dr. Carl Unger was supposed to talk about 'The Autonomy of Philosophical Consciousness and Nelson's Neo-Friesianism'. In the obscure journal mentioned by Nelson a certain Walter Johannes Stein had reviewed Kerler (1921). As a curious aside, this book by Kerler also contains a criticism of Nelson.

To enable you to understand Mr. Kerler's argument I must say something about the target of that argument. In his *Philosophy of Freedom* Rudolf Steiner teaches that concepts require reality, in other words that universals exist (a doctrine quite similar to the ones I have been discussing in these lectures in relation to other philosophers). Steiner's teachings play a role in his view of the self. He does not say that the self is a general object, I grant him that, he does not straightforwardly speak of the universal self, but he does speak of universal thinking as something that allows us to understand how the individual self is possible. He says that it is not the case that there is first a self which then thinks, but the other way around—there is first universal thinking which when directed towards itself generates a self. At least that is how Steiner's view is presented and interpreted in Mr. Stein's paper, and I quote[2]:

> One should not overlook the fact that it is thinking that allows us to define ourselves as subjects in contrast to objects... Thinking lies *beyond* subject and object. Thinking creates these two concepts the same way that it creates all concepts. If we as thinking subjects relate a concept to an object, we should not conceive of this relation as merely subjective. It is not the subject that brings about the relation; thinking does. The subject does not think because it is a subject, but it reveals itself to itself as a subject because it can think. The activity performed by man as a thinking being is therefore not merely subjective; it is rather neither subjective nor objective; it goes beyond both concepts. I should never say that my individual subject thinks, but my individual subject lives by the grace of thinking.

There is thus no such thing as a self that thinks, but thinking first creates the self, and this is so because it is thinking that creates the concept of SELF. Now this line of argument proves too much, for what is true of the concept of SELF is equally true of every concept. The concept of BEING is first created by thought; thus we must infer that there is nothing. This is analogous to the idea that there is no self prior to the creation of the concept of SELF by thinking. And the same thing would be true for thinking and the concept of THINKING.

But this is not the issue now. Mr. Kerler's argument is—to put it tersely, something Mr. Kerler does not do—that Steiner mistakes the self for the concept of SELF. Thinking creates the concept of SELF, but it does not create the self, as it does not create any other object whose concept we create by thinking. To put the argument in a nutshell: 'The self—or the subject—must have already existed before it came to apprehend itself. There must be a subject before it performs the act of knowing itself', and so on. This is a perfectly good argument, yet it is rejected by Mr. Stein. Let's see how this is done. I quote again[3]:

> Is the structural complex as *such* a subject, or does it become so only when it is *apprehended* as such? For this is what distinguishes a subject from everything else, e.g. a plant.

---

[2] Stein (1921, 3; it is a quotation of Steiner 1894, Chap. IV). The reader who finds all this rather quaint and dated is kindly reminded that quite similar arguments have been, and are still being, offered both by philosophers (e.g. by Heidegger and his followers) and social scientists (e.g. by a certain brand of Durkheimians). In fact, even certain analytic philosophers fond of cognitive externalism occasionally surf these waters.

[3] See Stein (1921, 3).

A plant is a plant, even if one does not apprehend it as such. Yet a subject is only a subject when it *knows* itself as such. One cannot *be* a subject without *knowing* oneself as a subject. This knowing-itself-as-a-subject is a constitutive factor of being a subject. Kerler's argument against Steiner shows that Kerler does not see that anything he says about thinking is only true insofar as one is thinking about objects. Yet if one thinks about thinking, if one does what is needed to apprehend one's own subject, then one is thinking in a *reality*. For in that case one *creates*, not just *contemplates*, something. One creates the subject. Everywhere else creating and knowing are distinct from each other. Yet in the case of the *self* they are identical. One creates oneself as an 'I' when one knows oneself as an 'I'. The Schoolmen distinguished between *ante rem*, *post rem*, and in re when talking about *universalia*. The original idea [the prototype] precedes the creation of something (*ante rem*). Knowing an already existing thing presupposes its existence (*post rem*). Creating and knowing the subject is in re, for what is otherwise separated—the *ante rem* and the *post rem* —coincide in this in re. In the case of the self the three *universalia* are unified. This is the *mysterium magnum*... When Kerler (1921, p. 260) says, "I object to Steiner that he identifies two different things, 'subject' and 'conceived subject'. Before any conceiving the subject was already there", I reply: Sure it was there, yet as a stage of its existence at which it should not be called subject. We do and should call something 'subject' only when it is *apprehended* as subject in a cognitive act. Only when thinking reflects upon itself in thinking about thinking [can we say that] a subject emerges *realiter* through the act—the *cognitive* act—of grasping itself as an 'I'. This cognition is here constitutive of the subject and the subject lives by the grace of thinking.

How does this illusory refutation of Mr. Kerler's argument succeed? By introducing a nominal definition, by postulating that the subject should only be called 'subject' if it thinks itself as a subject. This is done twice in the passage quoted:

1. This knowing-itself-as-a-subject is a constitutive factor of being a subject.
2. We do and should call something 'subject' only when it is *apprehended* as subject in a cognitive act.

There are two arguments against this. First of all, if the subject deserves that name, then what is it to know itself as? *As a subject*. But this is a concept which is already used in the definition:

3. Something is a SUBJECT if it APPREHENDS ITSELF AS A SUBJECT

This is a circular definition of the sort we have been discussing. The subject is defined via the concept of SUBJECT. If we substitute the definiens, we arrive at the definition:

4. Something is a SUBJECT if it APPREHENDS ITSELF AS THAT WHICH APPREHENDS ITSELF AS THAT...

and so on, without ever coming to know what something is to apprehend itself as in order to be called 'subject'.

But let us put this argument aside and assume it to be an infelicitous formulation. Let us assume Mr. Stein meant rather that something is a subject if it knows itself, so that he was trying to define 'subject' via the relationship to itself. The concept RELATING TO ITSELF does not require the concept of the SUBJECT. *Everything that thinks itself is a subject*. No circle in this concept, but again we are left with a mere nominal definition. We should call anything a subject if and only if it is something that thinks itself. And indeed anything at all—hence also the individual self—first becomes a subject on the strength of this definition by thinking itself. Fine, but from this what is supposed to follow does not follow. What is supposed to follow from the definition is that no individual self can exist, i.e. that no self that is different from the object opposed to it can exist, without this thinking-itself—and *that* does not follow. For *that* to follow we would have to replace the concept of INDIVIDUAL SELF with the nominally defined concept of SUBJECT. The proposition to be proved, viz that these two things [INDIVIDUAL SELF and THAT WHICH THINKS ITSELF] are identical, is hidden in that silent replacement of one concept by another. Thus, we are begging the question and the inference is based on a *quaternio terminorum*.

Let me give you another example that explains the apparent fruitfulness of the dogmatic way of doing philosophy. The illegitimate method used here finds a particularly productive application in a subject I was recently illustrating, viz the philosophy of law. The ways of the Schoolmen pervade this subject today almost as unrestrictedly as they pervaded all other philosophical subjects before Kant appeared on stage. Disputes among philosophers of law are, if we look closely, usually disputes over definitions. This fact deserves our attention before any other. How can reasonable people (we ask ourselves) debate over the definition of concepts? It makes no sense at all. You can define a concept any way you choose, so those are verbal disputes, purely terminological questions, about how anyone wants to use a word, what concept he wants to associate with that word. No factual statement is at stake. So I ask again: how can a dispute over words take place among reasonable people? The only way we can explain the fact that such disputes do take place is that when people are apparently concerned with a definition they are silently thinking about a factual statement. Behind the scenes there always lurks a proposition whose truth value depends on what concept we associate with a particular word or phrase contained in the proposition. People are interested in the definition only because of this hidden proposition. For it is the proposition, and not the concept, that constitutes the foundation of the science people are trying to erect. We have to find out what that proposition is and decide whether it is true or false without any regard as to how a concept is defined.

The most striking example for this way of doing legal philosophy is afforded by Rudolf Stammler, a legal scholar, philosopher of law and professor of legal studies at the University of Berlin, who wrote the famous book *The Theory of the Right Law* (1902). The principle of the right law put forward by Stammler is what he calls the 'social ideal'. He introduces the concept of SOCIAL IDEAL by means of a definition. The social ideal is defined as follows:

SOCIAL IDEAL = THE COMMUNITY OF MEN WILLING FREELY.

According to Stammler, this concept is the foundation of legal scholarship and the philosophy of law, i.e. the theory of the right law. This concept yields the criterion for the right law. To decide whether a given legal order or even only a single legal decision is right, we just compare it with the contents of Stammler's concept of SOCIAL IDEAL. The question now arises, what does 'people willing freely' mean? This [new] concept is defined by Stammler so: 'Willing freely' means willing what is objectively right. We perform the usual substitution and get the following [writing on the blackboard]:

SOCIAL IDEAL = THE COMMUNITY OF THE PEOPLE WHO WILL WHAT IS OBJECTIVELY RIGHT.

Fine, and what is OBJECTIVELY RIGHT? This is also defined by Stammler:

OBJECTIVELY RIGHT = WHAT AGREES WITH THE SOCIAL IDEAL.

Hence,

SOCIAL IDEAL = THE COMMUNITY OF PEOPLE WHO WILL WHAT AGREES WITH THE SOCIAL IDEAL.

The concept to be defined re-appears in the definiens, and so must be presupposed as defined elsewhere. We can of course, as in the above examples, perform a substitution and obtain this:

SOCIAL IDEAL = THE COMMUNITY OF PEOPLE WHO WILL THE COMMUNITY OF PEOPLE WHO WILL..., *and so forth* ad infinitum.

On the basis of such a concept there is no way we could ever decide whether a given legal decision is right, for the concept is empty. All that exists here are marks of white chalk on a blackboard or marks of black ink on the white pages of a book entitled *The Theory of the Right Law*. Yet Stammler endeavours to decide questions of law by means of that empty concept.

Let us look at one such legal question, one that illustrates well the persistence of the way of the Schoolmen in the field of the law. By 'the way of the Schoolmen' I mean the grotesque attempt to decide questions of law by means of definitions. My chosen example is the problem of matrimonial fidelity. Stammler sets out to prove that fidelity in marriage is required by the right law, i.e. that matrimonial infidelity violates the criterion of the right law. For Stammler this conclusion is just a matter of being consistent, of reasoning according to logic. It is a straightforward consequence of trying to avoid contradiction. Thus we should presume that it is only the incapacity for logical thought that seduces a person into marital unfaithfulness. How does Stammler manage to prove that matrimonial fidelity follows from logic? By means of a simple definition: by just defining marriage thus[4]:

MARRIAGE = COMMUNITY OF RECIPROCAL UNCONDITIONAL SURRENDERING BETWEEN TWO PEOPLE.

From this definition he argues that each marriage partner owes the other partner unconditional surrendering, otherwise their community is not a marriage. I quote[5]:

> In matrimony the thought should be embodied that under perfect surrendering of one's own person one recovers it in receiving an equal unconditional surrendering from one's partner...

---

[4]The word 'surrendering' (with its quasi-military connotations) very imperfectly translates German *Hingebung*, whose main connotations are (a) dedication to a cause, (b) unconditional love, (c) yielding to one's partner during the sexual act.

[5]The three passages quoted are, respectively, on pages 576, 577 and 578 of Stammler (1902).

Matrimonial fidelity thus results from the simple exercise of logical thought.

You see I was not exaggerating. If one partner surrenders and the other does not, then a contradiction results, in that

> a legal relationship should be maintained whose exercise is made impossible by one partner, given that it relies on reciprocal complete surrendering.

What has Stammler proved here? It has certainly been proved that a community which does not satisfy the requirement of matrimonial fidelity cannot be called 'marriage' according to Stammler's definition. All Stammler has proved is that such a community cannot be so called if we are to maintain his definition. This result has a purely verbal significance; it may perhaps be of interest to a linguistic scholar, but it is of no interest to the husband. Marriage is defined by the exclusion of infidelity. That exclusion concerns just the *name* 'marriage', whereas Stammler's argument does address the legal question we wanted to decide, viz whether matrimonial fidelity is allowed or not, whether it is legally right or not.

Let us try to clarify the point by rephrasing our original question: should the community of the sexes be a marriage in the sense posited by this definition of marriage? In this form our original question re-emerges unanswered, for the proposition we want to decide was whether the community of the sexes should be a marriage. This proposition is not in the least addressed by Stammler's proof. What is sophistic about it is that it would plainly prove too much, even if in other ways there is nothing to be said against the proof. I hinted before that according to this proof matrimonial infidelity would be impossible—it would imply a logical contradiction. One could come up with a remedy to this deficiency by putting into the definition the idea that

MARRIAGE = COMMUNITY IN WHICH THERE *OUGHT TO BE* UNCONDITIONAL RECIPROCAL SURRENDERING.

But this manoeuvre does not help at all, for the concept of MARRIAGE that is defined by this requirement would be entirely inapplicable. In order to use this definition to establish whether a community ought to meet the requirement of fidelity or not we would have to know in advance that the community *is* a marriage. But in order to know this we should have already discovered (by means of the modified definition) that the community ought to be of the unconditional-reciprocal-surrendering kind, or of the fidelity kind, which was precisely the legal question we were supposed to decide by applying the definition.

Consider now another example drawn from the study of law whose importance is greater and more transparent.[6] I am choosing these examples deliberately so as to convince you that these are not just logical subtleties but rather philosophical fallacies with far-reaching practical consequences. I want to talk about the question of the League of Nations. Until around the end of the war students of law and philosophers of law, at least in Germany, held it as irrefutably true that there is no

---

[6]The whole question of sovereignty, which occupied and still occupies pride of place in political philosophy, is treated extensively in Nelson (1917).

such thing as a legal ideal of a League of Nations. The underlying argument relied on the celebrated concept of SOVEREIGNTY, widely considered to belong to the concept of STATE. According to this view the state would give up its very existence if it had to subject itself to a League of Nations, i.e. to a higher-ranking legal community. The state would lose its sovereignty, the independent authority that it necessarily possesses as a state, it would cease to exist as a state, it would abrogate itself. Accordingly, the idea of a League of Nations seems to be a self-contradictory concept, something as logically impossible as matrimonial infidelity is for Stammler. The fallacy people commit here comes from taking the concept of STATE, a concept that is given *before* any arbitrary definition, and replacing it with a new concept that is created by an arbitrary nominal definition. The new concept is indeed defined via the attribute of SOVEREIGNTY, i.e. LEGAL INDEPENDENCE FROM ANY HIGHER-RANKING LEGAL ORGANISATION. Armed with this nominal definition people falsely argue that the subordination of states under a higher-ranking organisation that legally regulates their mutual relations is in fact equivalent to their self-annihilation. To be sure, if the concept of STATE is defined through the attribute of SOVEREIGNTY, then the so defined state would be abrogated by the introduction of a League of Nations, but this has nothing to do with really existing states [as opposed to merely defined ones].

References

Kerler, Dietrich Heinrich. 1921. *Die auferstandene Metaphysik: eine Abrechnung* [*Metaphysics resurrected: A settling of scores*]. Ulm: Verlag von Heinrich Kerler.
Nelson, Leonard. 1917. *Die Rechtswissenschaft ohne Recht: kritische Betrachtungen über die Grundlagen des Staats- und Völkerrechts, insbesondere über die Lehre von der Souveranität* [*Law without justice: Critical reflections on the foundations of public and international law, in particular on the theory of sovereignty*]. Leipzig: Veit & Comp [Reprinted in Nelson (1971–1977), vol. IX, 123–324].
Nelson, Leonard. 1971–1977. *Gesammelte Schriften*, 9 vols. eds. Paul Bernays, Willy Eichler, Arnold Gysin, Gustav Heckmann, Grete Henry-Hermann, Fritz von Hippel, Stephan Körner, Werner Kroebel, and Gerhard Weisser. Hamburg: Felix Meiner.
Stammler, Rudolf. 1902. *Die Lehre von dem richtigen Rechte*. Berlin: J. Guttentag. [English translation: *The theory of justice*. New York: Macmillan (1925)].
Stein, Walter Johannes. 1921. Zur Verteidigung der 'Philosophie der Freiheit' [In defense of the 'philosophy of freedom']. *Dreigliederung des sozialen Organismus* 3: 2–3.
Steiner, Rudolf. 1894. *Die Philosophie der Freiheit: Grundzüge einer modernen Weltanschauung*. Berlin: Emil Ferber. [English translation: *Philosophy of freedom: A modern philosophy of life developed by scientific methods*. London: G.P. Putnam's Sons (1916)].

Lecture XVI

Abstract The concept-swapping fallacy could be dismissed out of hand as a mere playing on words that does not deceive anyone were it not for the fact that under certain circumstances it can, and does, have serious practical consequences. This is eminently the case of the alleged proof that a supranational organisation such as the League of Nations is a logical impossibility, for it would destroy the nation-states that would try to set it up. This 'proof', constructed of all pieces by German legal scholars, was instrumental in preventing the establishment of such an organisation and in preparing the ground for future war.

I had just begun to expound an example to show you that avoiding typical philosophical fallacies is not a matter of logic-chopping. Some of these fallacies may have consequences of highly significant and practical import, far beyond the fate of individuals, for they may affect the fate of entire nations and regions if not indeed the fate of all of humankind. You will soon see that I am not exaggerating, for my topic is the League of Nations and my aim to show you that confusion as to the relevant legal concepts in this subject causes the most immense practical confusion, which can understandably bring chaos into the affairs and future of nations. I would like to uncover again the root of the fallacy here.

The concept of STATE is one of the concepts with which we are acquainted before any definition. It is a familiar concept for all of us and especially for all students of law, all the more so if they are experts in public and international law. They must use it, and they do use it, before any definition. This explains the interest raised by the disputes about how to define the concept of STATE. I mentioned earlier that it is something of a puzzle that reasonable people can quarrel over a definition. This puzzle can only be solved if we remember that we possess some concepts before defining them, so that whoever tries to produce a definition, even if he is not clear in his mind that any definition he may propose should not be arbitrary, he will nonetheless have at least an obscure feeling that he is not free to propose any definition he likes, that somehow there are limits to any definition he might want to propose. And that is why people take an interest in the right definition. A concept does not in itself assert anything. It can in itself be defined any way one likes—if it were not the case that it *is* unconsciously associated with one assertion, viz that the proposed definition is actually equivalent to the concept originally given so that the concept as defined contains exactly the same attributes as the concept we had before the definition. The claim is, whether one admits to it or not, that these two concepts have the same content or at least the same extension, that they refer to the same objects.

It is *that* hidden claim that is important; it is the only matter in dispute, albeit the disputants are in general not clear in their minds that this is the case. We found something similar in the earlier example of the concept of MARRIAGE. We also have a concept of MARRIAGE before we introduce a definition for it, and we will all take for granted (unless it is explicitly denied) that the intention of whoever introduces the definition is to define the very concept that we had of marriage *before* the definition was produced. After all, we have no reason to believe that he would want to pull the wool over our eyes and play games with words. But in fact if somebody produces a definition whilst he is not clear in his mind about all this, then he is exactly like one who is playing word games and pulling the wool over the eyes of his readers, listeners and even his own. For he has not paid attention to the task at hand, viz to define the original concept and not another one, so that he will be smuggling in an assertion to the effect that the concept as he defines it and the original concept have the same extension. His later claim to establish that assertion on the strength of his definition is thus not in the least warranted.

It is similar in the case of the concept of STATE. Experts in public and international law, primarily German ones, are used to defining the concept of STATE by means of the attribute

BEING INDEPENDENT FROM ANY HIGHER AUTHORITY,

so they have an easy time proving that the ideal of a League of Nations is not only utopian but self-contradictory, an altogether impossible thing, and they treat as foolish all those who work hard at making it real. They even went so far as to say—an easy thing given their definition—that a state trying to join a League of Nations would abrogate itself. To establish a League of Nations would be a crushing attack on the existence of individual states.

It is well-known how much ground this view has gained, not only among large numbers of legal scholars, but how much it has actually influenced the run of events and the destiny of nations. You will be aware of the fact that at least the first step towards a League of Nations has been taken, viz the Permanent Court of Arbitration of the Hague Conventions. The majority of civilised nations were in favour of taking this first step towards a League of Nations. The attempts were thwarted by the invincible resistance of the German Reich.

The German Reich had sent Mr. Stengel as one of its representatives to The Hague Conferences (in 1899 and 1907). Now Mr. Stengel is one of our experts in public and international law. He has written quite a few books, including *World Government and the Peace Problem* (1909), in which he explains that only states, i.e. sovereign polities, can be members of the international community.[1] If one assumes states to be independent polities not subject to a higher authority, it follows

---

[1] See Stengel (1909, 4).

that the sovereignty of states is and must be the foundation of international law. There would be no need for international law in Mr. Stengel's sense of the word if, as is the aspiration of pacifists, we should already have brought the states of the international community together into a world state. The relationship between the states would be regulated by worldwide public law.[2]

What is going on here? Mr. Stengel has produced a definition of the concept of INTERNATIONAL LAW, according to which[3]

INTERNATIONAL LAW = THE SET OF ALL LEGAL RELATIONS BETWEEN SOVEREIGN STATES

It is therefore clear that the elimination of state sovereignty and the establishment of the higher-ranking League of Nations are inconsistent with international law *so defined*. To establish a confederation of states would amount to a violation of international law *so defined*. According to the concept of INTERNATIONAL LAW as Mr. Stengel defines it, international law *requires* the preservation of the states united by it and *forbids* the League of Nations. This is the kind of wordplay I was talking about. The wordplay here concerns the phrase 'international law'. Everyone is free to define the concept that they wish to associate with this phrase. But they will have to be aware of the fact that the original concept of INTERNATIONAL LAW, already in use before any definition, is not thereby eliminated. They must therefore prove what is the interest (for us or for the states) in preserving international law *so defined*. This proof is entirely lacking. The only consequence is that establishing a League of Nations amounts to eliminating international law, a proposition that can shock people who are not used to thinking but will not impress anyone who does not allow himself to be led up the garden path by a pun.

As I said, the original concept of INTERNATIONAL LAW is not in the least affected by this, and hence international law can exist after it was defined by Mr. Stengel just as well as it existed before, perhaps even better than it did before. By contrast, someone who overlooks the distinction between the original concept and Mr. Stengel's concept will labour under the illusion that something has actually been proved and that there is a legal impediment to the League of Nations. However, it has not even been proved that Mr. Stengel's concept is real, i.e. that there is such a thing as international law as defined by him, let alone that his injunction agrees with a legal norm. No proof at all has been given, and if we look more closely, not only is Mr. Stengel's conclusion unproven but one can see that it is false. Real international law, according to the original concept, is not inconsistent with the League of Nations but actually requires it, and the same is true of the state. The STATE defined by Mr. Stengel via the attribute of SOVEREIGNTY is certainly abrogated by establishing a League of Nations. But why should it not be abrogated?

The illusion comes from using the word 'state' to refer to the entities that we call by that name—and some people tremble in fear that the state in which they live might perish. But we are not talking about these states. The attribute of SOVEREIGNTY

---

[2]See Stengel (1909, 93, 94).

[3]See Stengel (1909, 1).

does not define these states at all. I do not want to put forward a definition here, but this much is clear: that we mean communities of people which have their own government, their own laws, their own jurisdiction, their own administration. To be sure, something of the sort happens: laws are made for the individuals of a nation by its government independently of a higher-ranking power. But if we ask ourselves whether the existence of this kind of law would be endangered, threatened or even abrogated by becoming part of a confederation of states, then we need only remind ourselves that such a confederation will simply regulate the mutual relations and dealings between the states yet for the rest will not intervene in the internal relations of any state, in its internal questions of legislation, judicature or administration. Indeed we discover that the existence of the state is not in fact threatened by a League of Nations, but quite the opposite, it is protected by it. Just as the individual in the state would be unprotected and exposed to predatory attacks by his fellow citizens if it were not by the government, the laws, the courts, the police, in the same way the state is exposed to predatory attacks by its neighbours as long as there is no superordinate power above the states that legally regulates the relation between the individual states and safeguards the rights of the one against attacks by the other. It is quite out of the question that the League of Nations would be an attack on the continued existence of states, since it is in fact directed towards their preservation. This is the exact opposite of the deceptive ideas created by fallacious argumentation. The deception vanishes as soon as we get rid of the arbitrary nominal definition.

However, this is not about one particular author who as a matter of fact happens to have played a calamitous political role. This is rather about a whole way of thinking that was generally current in legal studies, at least in the German-speaking world, up until the end of the war. There is no better way to prove that than by looking at Georg Jellinek, the most famous scholar of public law in modern Germany if not the most famous in contemporary Europe. The writer of the *General Theory of the State* (1914), which all legal scholars have read, holds the following view both in that book and in his other books. His arguments proceed always from the definition of the concept of SOVEREIGNTY. This fact is already worth our attention. For what would be the interest in such a definition unless the concept of SOVEREIGNTY hides the silent and always implicit assumption that it refers to something real, that there is such a thing as sovereignty, i.e. a right to sovereignty? [Jellinek says:][4]

> Sovereignty is the right to enter obligations by one's own will.
> [Sovereignty is] the attribute of a state by which it can be legally bound by its own will.

---

[4] The first quotation is on p. 55, the second on p. 34 of Jellinek (1882). The German here translated by '*one's* own will' and '*its* own will' actually lacks expression of the *possessor* or *subject* of the will, which in English is impossible. This is part of the linguistic sleight of hand perpetrated by Jellinek. The question of *whose* will we are talking about is thus much harder to raise for the original German definitions. The state as such cannot be said to *will*, only certain individuals or groups who are in a position to make the relevant decisions.

The question of whether there is such a thing as a right corresponding to this definition does not arise at all, and it does not because it is considered self-evident that there *is* such a thing. Anyone who makes the effort actually to raise this question will almost by the very act of raising it give a negative answer. The right as defined is a contradiction in terms, and we do not need to inquire any further as to whether that right exists. We know in advance that it cannot exist, for it is as inconsistent as the underlying concept of INTERNATIONAL LAW. For the SOVEREIGNTY of one state is in direct contradiction with the SOVEREIGNTY of any other state. SOVEREIGNTY would be the lack of any legal limits to state activity. Now the legal lack of restriction of one state is made possible only by restricting the rights of any other state. A state can only be sovereign relative to another state if the latter has no right before the former. Otherwise the rights of the latter would infringe upon the freedom of the former and thus be equivalent to a destruction of its sovereignty.

The usual way of wriggling out of this difficulty involves a sophistical reply, viz that it is precisely the sovereignty of the states which makes it possible to establish legal relations between them. It is by state sovereignty that the state restricts itself in its dealings with other states—by signing treaties. This reply sounds quite plausible. But if we delve more deeply into the question, then the contradiction indicated above rears its head again and becomes even clearer. To begin with, there must certainly be a right of states to make treaties, i.e. agreements about their dealings with each other. The right of one state to make an agreement with any other state would already nullify its sovereignty, for every right of a state is based on a restriction of another state's right, and therefore nullifies its sovereignty. Secondly, what would be the basis for the obligation to fulfill a treaty if we derive the legal status of a treaty from state sovereignty? If a state is sovereign, then it is bound by nothing other than its own will—whether it will stick to the agreement or breach it is only for the state to decide.[5] And one cannot ground the obligation to be faithful to the agreement on a further agreement that creates an obligation to respect all agreements made between states. For the above argument would apply to that further agreement. In order to bind the states to that further agreement a previous one should be presupposed which contains the obligation to respect this third agreement, and the same would be true of *that* agreement, and so on and so forth ad infinitum. All this is nothing but sophistry and legal wrangling.

This much can be said about the concept of SOVEREIGNTY; but the same applies to the concept of STATE. Jellinek defines STATE as follows[6]:

> Any ORGANISED SECULAR COMMUNITY WHICH IS NOT SUBORDINATED TO ANY ORGANISATIONAL STRUCTURE, is a STATE.

What was supposed to be proved follows immediately from such a definition, viz that the concept of an ORGANISATION OF STATES is just a utopian ideal, and not only

---

[5]It is interesting that Nelson, usually quite sharp, in this passage seems to consider the State as a subject of will. See Footnote 4 in this chapter.

[6]See Jellinek (1914, 365).

that, but also that it is a contradiction in terms and thus logically impossible. Which is exactly what Jellinek says[7]:

> This concept [CORPORATION OF STATES] is self-contradictory and so cannot be realised.

It is not just that physical hindrances, legal worries, considerations of expediency or such things stand in the way of establishing a League of Nations—no, it is *logic* that forbids it.

What reinforces this sophistry is the historical fact that what we call 'states' today agree with Jellinek's definition—they are not subordinated to a higher-ranking organisational structure, if we leave aside some difficult cases of states incorporated into federations.[8] Yet this historical fact depends on there not being any League of Nations. For that is precisely what the nonexistence of the League of Nations means—that these states have no organisational structure above them. It is therefore historically correct to say that the states in the currently idiomatic sense of the word do fall under the concept as defined by Jellinek, yet this historical matter of fact is mistaken for a legal and even a logically necessary state of affairs.

We can deepen our grasp of the logical trick, for Jellinek explains in detail his idea of what is the purpose of a definition. It is very instructive. He says[9] that legal scholars come to create a concept

> by careful examination of all phenomena which fall under the concept and by comparing and bringing together their common attributes. It is then a matter of induction—legal concepts are as inductive as any other concept abstracted from experience.

So the way to arrive at a definition of the concept of STATE is for Jellinek a two-step procedure. The first step is to check which attributes are *as a matter of fact* shared by the existing states. The second step is to bring them together. This is how we obtain the definition of the concept of STATE. In Chapter "Lecture IV" I explained that not all attributes shared by a set of objects belong to the content of the corresponding concept. A concept that would be so defined would be logically over-determined. To the concept belong those and only those attributes shared by a set of objects which are necessary and sufficient for an unambiguous specification of those objects, i.e. which are necessary and sufficient for deciding whether a given object belongs to the set or not. Adding any further common attributes beyond those contained in the concept can only be done by a synthetic judgment—and indeed a synthetic judgment a posteriori if it concerns historical phenomena.

It is entirely possible (as I said in Chapter "Lecture IV") to create a concept by putting together all attributes common to a class of objects. One should realise, however, that such a concept is different from the one that only contains the necessary and sufficient attributes of the class. They have the same extension but

---

[7]See Jellinek (1914, 772).

[8]Nelson refers to cases such as the United States of America, born as a federation in 1776, as well as the newly unified states of Italy (1870) and the German Reich (1871).

[9]See Jellinek (1882, 11).

not the same content. The judgment that asserts that the objects that fall under the concept with less content also fall under the concept with more content, is a synthetic judgment. Even if one combines all attributes common to the elements of a set, the co-extensionality of the two concepts cannot be obtained by logical analysis or comparison of the concepts. It may be argued that Jellinek's idea of arriving at a definition by induction, i.e. by finding out all the attributes common to the objects of a class, is justified by the desire to get ahead of experience by sheer speculation. Yet it is this procedure that unavoidably gives rise to all those speculative encroachments that occur when people pass off coincidental, purely empirical facts as necessary consequences of a definition. A misleading game is played here with the concept of STATE, whereby the conclusion is drawn [to repeat the quotation] that

> [the] concept [of a CORPORATION OF STATES] is self-contradictory and so cannot be realised.

The only actual result of these verbal manipulations is just that the subjection of states to a League of Nations has as a consequence that the political entities one has in mind should no longer be called 'states', for now only the League of Nations would be a STATE as defined by Jellinek, viz an ORGANISATION THAT IS NOT SUBJECTED TO A HIGHER LEGAL ORDER, whilst the former states are now so subjected, and so can no longer be called 'states'. But it is a great puzzle why we should consider it a misfortune either for the states or for the world or for humankind, if this name could no longer be used in the sense those authors love so much. This fallacious doctrine of the sovereignty of states must be traced back to Hegel (1821), and all these fallacies are already present in Hegel's philosophy of law.

One cannot dodge this criticism by taking the requirement that the states should be sovereign and putting it into the definition of STATE, as opposed to a definition in which the state is an entity that is sovereign as a matter of fact. The fallacy cannot be avoided by defining the STATE via the condition not so much that it *is* but that it *should be* independent of any higher power, because to apply this [normative] definition to a given political entity we would have to know beforehand that it *is* a state. Otherwise it would not be possible to use the definition as a legal postulate from which to extract a conclusion about the entity to which the concept of STATE is applied. In order to subsume that entity under the concept of STATE, we would have to establish beforehand that the object actually possesses the defining attributes contained in the concept, i.e. we would have to know beforehand that the object should be sovereign, which is the proposition that should be proved.

References

Hegel, Georg Friedrich Wilhelm. 1821. *Grundlinien der Philosophie des Rechts*. Berlin: Nicolai.
 [English translation: *Elements of the philosophy of right*, Cambridge University Press, 1991].
Jellinek, Georg. 1882. *Die Lehre von den Staatenverbindungen* [*The theory of associations of states*]. Vienna: Alfred Hölder.

Jellinek, Georg. 1914. *Allgemeine Staatslehre* [*General theory of the state*], 3rd ed. Berlin: Oscar Häring [First edition 1900, second edition 1905].

Stengel, Karl Michael Joseph Leopold, Freiherr von. 1909. *Weltsaat und Friedensproblem* [*World government and the peace problem*]. Berlin: Reichl & Co.

Lecture XVII

Abstract The fallacy used to declare the League of Nations inadmissible under the pretense that it is incompatible with the concept of the state can be analysed as the joint work of three nominal definitions. If instead of states we would take individuals, this method would produce the strange result that an institution to protect individuals is proscribed by law. It is shown that the fallacy is present in the writing of several eminent legal scholars. A similar fallacy sometimes also appears in a theoretical field such as arithmetic, where it has been used by some philosophers such as Leibniz and mathematicians such as Grassmann, Peano and Poincaré to prove that arithmetic theorems are analytically true.

In our last session I was using examples taken from distinguished legal scholars to elucidate the role played by logicistic metaphysics in this field even today—the staging of an illusion that knowledge can be extended by producing arbitrary nominal definitions to enable a surreptitious transformation of an analytic judgment into a synthetic one. My examples were chosen from the legal field not only because this fallacy is least contested and most widely held in esteem there, but also because it is there that you can see the fallacy not just as a matter of logic-shopping (interesting to academics, logicians, philosophers or at most legal scholars), but as pregnant with hidden and all the more insidious practical consequences, affecting the fate of individuals, nations, and perhaps ultimately of all humankind. It is tempting to develop those consequences here, but we must stay put and go on with the theory of this logical fallacy, leaving to you the task of extracting the practical implications. If you did, you might write a new and so far neglected chapter in that controversial book that people are writing as we speak about who was responsible for the last war. For it is real, flesh-and-blood people who have to pay for the tragic consequences of the fallacies committed by their intellectual leaders.[1]

We have been examining the idea of a League of Nations, i.e. the idea of an organised legal order between states. The logical fallacy here comes from an arbitrary definition, first of the concept of INTERNATIONAL LAW. This is the starting definition:

---

[1] The idea that World War I was a failure of thinking and reasoning (and not a failure 'of the heart') is not exclusive to Nelson, as can be seen in Keynes (1919) and Collingwood (1939).

(1) *First Nominal Definition:* INTERNATIONAL LAW = REGULATION OF INTER-STATE RELATIONS.

Legal scholars add more details about 'regulation' and 'relations', but we are here more interested in the word 'state'. And so we get to our

(2) *Second Nominal Definition:* STATE = ORGANISED COMMUNITY WHICH (as legal scholars say) IS SOVEREIGN (i.e. not a member of a superordinate community or body politic, in brief a community that is regulated independently of any other authority).

The crucial concept here is SOVEREIGNTY, i.e. INDEPENDENCE FROM A SUPERORDINATE BODY POLITIC. The state is then a body politic that is not a member of a higher-ranking body politic. At this point legal scholars come to the problem of a League of Nations and, applying their sharp legal wits, introduce the term 'state union' ['state confederation'] thus yielding our

(3) *Third Nominal Definition:* LEAGUE OF NATIONS = ORGANISED COMMUNITY OF STATES.

The key word in this definition is 'organised'. It means as much as the phrase 'body politic'. Organised is the opposite of 'anarchic', i.e. relying on purely individual voluntary agreements. A community is organised if there are authorities that regulate inter-individual relations, authorities to which the individual members of the community are subordinated. Now it appears that one merely needs to put these definitions together to arrive at the following conclusion:

(4) *Conclusion:* Establishing a League of Nations is incompatible with international law.

We see that the two concepts of INTERNATIONAL LAW and LEAGUE OF NATIONS are opposed. International law would proscribe any League of Nations. But wait, what is it exactly that international law would proscribe? The establishment of an organisation of states to uphold the law between them, for that is how the League of Nations is defined. A paradoxical consequence then, obtained by a trick. The trick becomes plain the moment we substitute 'private law' for 'international law' and 'individuals' for 'states'. We obtain the following argument:

> Private law is the regulation of inter-individual relations
> Individuals are sovereign
> The state is an organised community of individuals
> ———
> Therefore, establishing a state is incompatible with private law

In other words: anarchy in the dealings between single individuals is commanded by law. Establishing an organisation that upholds the law in all dealings between individuals is incompatible with the law. *Anarchism is apparently a logically necessary legal principle.*

What's happened here? Something is not quite right. In conclusion (4)—to repeat, 'establishing the League of Nations is incompatible with international law'—the phrase 'international law' is taken to be the law regulating inter-state relations, the set of all legal relations between states, and the word 'state' means what every unbiased person understands by it, viz an organised community of individuals, a community of people such that the law applying to them is upheld by a higher authority. But note: a paradox emerges only if we take the word 'state' in this ordinary meaning. If we actually assume the definition given, there is nothing paradoxical about the conclusion. But in that case conclusion (4) says nothing that could be of any practical interest to us. What touches us from a practical standpoint is rather the fate of the state in which we live, i.e. the organised community of individuals to which we belong. So, if by 'international law' we do understand the set of all legal relations between the states, as we actually conceive the latter [without the misleading attribute of SOVEREIGNTY], then conclusion (4) does not follow, indeed the League of Nations no longer appears to be incompatible with international law, but precisely the opposite—we see it as postulated by international law, just as the state is postulated by inter-individual law. It is the same here as there: Just as the state upholds the laws that protect the individual against violence and despotism on the part of his neighbours, so will a legal order ranking above the states—a League of Nations in fact—uphold the law that protects any given state against despotism and violence exercised by its neighbours. The fallacy was to take the attributes arbitrarily attached (by a nominal definition) to the concepts of STATE and INTERNATIONAL LAW and to transfer them to the ordinary and very different concepts that go by the same names. It is only by confusing the concepts introduced by fiat with the original concepts that the advocates of an international rule of force are able to champion despotism and violence in the name of the law and to decry the introduction of a legal inter-state order as a state-jeopardising attempt, a manoeuvre that has unfortunately found compliant believers among gullible nations.

Although I have given you a detailed examination of the views of such a universally regarded scholar of public and international law as Jellinek, someone might still think that—given just how appalling the line of argument involved is—those views are exceptional, and harbour doubts as to whether we are really dealing here with a general fallacy. So I would like to read a few excerpts from several other famous professors of international law. I quote first from Professor Ernst Zitelmann[2]:

> International law is not enforceable and it *cannot* be, for any true state is sovereign, *ie* not controlled by any state power but its own; if it were subject to a higher state organisation endowed with legitimate authority, then it would not be a sovereign state any more, but rather a member of a new state, a confederation, and we would not have to do with international law but at most with federal public law.

---

[2]Zitelmann (1914, 477).

It looks as though the existence of an individual state that is incorporated into a League of Nations, is entirely abrogated, absorbed by a world state. This would certainly be the case if we define STATE in the proposed way, if we say that a true state *is* sovereign. The word 'true' is questionable, for what is a true state? A political entity is either a state or not a state. Whether it is a *true* state, i.e. one agreeable to the author's sentiments, that is another matter altogether, and the author should first prove that a state is also a true state, i.e. one he likes; but that is not what we are talking about and the fallacious conclusion has been smuggled in.

Another quite well respected scholar of international law, Professor Paul Heilborn, says[3]:

> [A] state should not sacrifice itself for other states, but in all cases of conflict it should think only of its own welfare. This is a truism we have to repeat again and again.

There! That one state does not have to think about the rights of other states, is a truism for an international law scholar! From there it is not at all difficult to argue that international law is incompatible with a legal organisation of states[4]:

> The concept of JURISTIC PERSONALITY legally implies that of STATE...

The League of Nations would be one such juristic personality, and so legally a state, and because a state is sovereign, there would be no other states in the world but only the one true state, called 'League of Nations':

> The concept of JURISTIC PERSONALITY legally implies that of STATE, and its law would regulate the relation between the union and its members, in exactly the same way we know from confederations. International law would be no more.

International law would be no more *if* that was the only way to uphold it.

This whole view has also gained a foothold here in the Law School of the University of Göttingen. Our very own international law scholar, Professor Paul Schoen, says about the topic of the objective law between states[5]:

> Who should establish that law in a way that binds sovereign states?

An unbiased person who tackles this question would immediately think that establishing that law concerns those who have the expertise to find out the truth, and because the truth we are looking for belongs to the law, and specifically to international law, finding it out should be left to legal scholars and specifically to international law scholars. But that is not Professor Schoen's answer:

> Neither the views of this or that legal scholar nor the views of the majority of legal scholars can be a standard for states to meet.

---

[3]Heilborn (1912, 21).
[4]The two following quotations are from Heilborn (1912, 35).
[5]The three following quotations are taken from Schoen (1915, 290).

Lecture XVII   155

Mark well: Professor Schoen is saying that what legal scholars find to be the legal truth by following their scholarly methods yields no standard for states to meet. Instead:

> The states themselves have the task of examining the value of this or that norm, and even if they are convinced that a given norm is appropriate, they are under no formal obligation to follow it. For the very concept of a SOVEREIGN STATE implies that it is not bound by any norm that it does not want to be bound by. Even a norm that states have objectively examined and recognised to be the appropriate norm can be formally and legally rejected by them as a binding one if for whatever reason they do not want to be bound by it.

Of course, it is undeniable that if 'sovereignty' means 'the independence of the state from all legal norms that it does not want to follow', then the sovereign state cannot be bound by such norms. But what follows from this? As to real states nothing at all follows, unless it is first proved that they *should* be sovereign. But whoever raises this very question—should they be sovereign, is there such a thing as a right of sovereignty?—will find at once that the right of sovereignty contradicts itself. This was covered in our last lecture. The legal possibility of a sovereign state is inconsistent. The concept of SOVEREIGN is defined as INDEPENDENT FROM ANY LEGAL NORM. Every right of any state, no matter which we take but first of all the right of sovereignty, would abrogate the sovereignty of every other state, for the rights of one is only possible by restricting the rights of the other. Sovereignty, as a universal right of states, is therefore self-contradictory.

One last international law scholar, Professor Erich Kaufmann, deserves to be mentioned, because his effort really crowns this whole perversion of the law. Kaufmann's book *The Essence of International Law* (1911) enjoys so much recognition that he was offered no less than the Chair of International Law at the University of Berlin. The principle he preaches is that because of state sovereignty the social ideal is war.[6] The dream of this German scholar has in the meantime been fulfilled.[7] Whether he relishes the actual outcome of the war [for Germany] is of course another matter.

No more needs to be said about this typical fallacy in the field of law. However, the fallacy is so ingrained that I would like to consider two further examples from another field, in fact from a field where one would least expect to find this fallacy and its alarming consequences, viz in mathematics, specifically in that branch of mathematics where philosophy is at home, the question being not the truth of mathematical theorems (as decided via the usual mathematical methods) but rather their origin and their axiomatic presuppositions. The case of geometry was covered in earlier lectures, so I want to show now that it is the same in arithmetic.

I have already said that an axiomatically rigorous treatment of arithmetic—of the sort we have had, from Euclid on, for the case of geometry—was only achieved in the nineteenth century. The problem of axiomatic arithmetic itself was raised earlier. Leibniz in particular tried his hand at solving it. He maintained (as we saw in

---

[6]See Kaufmann (1911, 146).
[7]Nelson refers to World War I (1914–1917).

Chapter "Lecture IV") a logicist view as to the origin of arithmetic truths. And so he believed the theorems of arithmetic could be proved via pure logic by means of mere definitions. He offers a simple example himself. It concerns the proposition $2 + 2 = 4$. Leibniz gives the following definitions to set up the proof.[8]

(1) $2 = 1 + 1$
(2) $3 = 2 + 1$
(3) $4 = 3 + 1$

The proof is carried out as follows:

(4) $2 + 2 = 2 + 1 + 1$ according to Definition (1)
(5) $2 + 1 + 1 = 3 + 1$ according to Definition (2)
(6) $3 + 1 = 4$ according to Definition (3)

therefore,

(7) $2 + 2 = 4$ which was to be proved

This proof seems complete and without a gap, as though it had been carried out solely on the basis of the definitions given as starting-points. If one looks more closely, however, it is clear that the proof only works because the brackets (whose necessity we appreciate nowadays) are left out, particularly in step (5). If things were above board, one would have to write (5) as follows:

(5′) $2 + (1 + 1) = (2 + 1) + 1$

and one could then via definitions (1)–(3) argue that

$2 + 2 = 3 + 1 = 4$.

We thus see that we are surreptitiously using a presupposition that is independent of the definitions given, which has the general form of the associative law of addition

(A) $a + (b + 1) = (a + b) + 1$.

Whether (A) can be proved by purely logical means is too deep a question to tackle here and in any case irrelevant for our purposes. What is relevant for us is that a particular presupposition is made here which is not contained in definitions (1)–(3) that were given as starting-points, viz (A), which, more to the point, cannot be introduced in the proof as a mere definition. People have indeed tried to make the associative law into a definition—the definition of the concept of ADDITION—in

---

[8]The proof that follows is to be found in Leibniz (1704, Book I, Chap. VII, §10). Nelson's critique probably stems from Frege (1884, Chap. I, §6). It is a bit curious that Nelson does not mention Frege's name, either here or in the other passages of his works in which he reproduces his refutation, although he was demonstrably aware of Frege's paternity. It is in fact all the more curious given that Frege's treatment of the question is far more sophisticated than Nelson's. In particular, Frege also has '$2 + 2 = 2 + (1 + 1)$' and the general form of the associative law '$a + (b + c) = (a + b) + c$'.

Lecture XVII

order to improve on Leibniz's proof, and they keep trying. It *is* very tempting to do it. Hermann Grassmann, a profound mathematician, tried to do it; Giuseppe Peano, highly deserving of credit for his axiomatic treatment of mathematics, tried to do it; even Henri Poincaré in our own days has tried to do it.[9]

Let us try to see why the associative law (A) is not a mere definition, i.e. that the left side of equation (A) has its own meaning, independent of the meaning of the right side. If (A) were a definition, then the meaning of the symbols on the left would be defined by the symbols on the right. Everyone is free to choose what meaning he wants to assign to the symbols he uses; and so it would be all right for him to say that the meaning of what is on the left of (A) is the same as the meaning of what is on the right. We might then ask why does he bother to introduce two series of symbols for one and the same thought, where in fact the second series is not even shorter than the first. However, the question is whether the scientific content of arithmetic is preserved by this procedure. Can even such a simple operation as the addition of 2 and 2 be carried out within this kind of arithmetic? No, it cannot. The problem of adding 2 plus 2 cannot be raised, let alone solved, in this arithmetic. The problem of how much 2 plus 2 makes cannot be settled. For say we write the following definition [as a special case of the law (A)]:

(8) $\quad 2 + (1 + 1) = (2 + 1) + 1$

In (8) the sequence of symbols on the right explaining the meaning of '$2 + (1 + 1)$' [Nelson writes '$2 + 2$'] would be '$(2 + 1) + 1$'. The symbols '$2 + (1 + 1)$' would mean nothing else. [In particular, they could not mean '$2 + 2$'.][10]

Within the arithmetic given by definition (8) plus definitions (1)–(3), we can certainly raise and solve problem (9):

(9) $\quad$ Is $3 + 1 = 4$?

for the expression '$3 + 1$' is defined as '4'; but there is no way to raise or solve problem (10):

(10) $\quad$ Is $2 + 2 = 4$?

The original problem, the one Leibniz wanted to solve by the procedure he proposed, has been stolen away.

Let us make this more intuitive. Imagine we have two baskets with apples, one on the left and one on the right. In the left-hand basket there are a apples, in the right-hand basket there are b apples. Consider instruction (11):

(11) $\quad$ You put first an extra apple into the right-hand basket [$b + 1$]. Then you put the apples from the right-hand basket into the one on the left [$(b + 1) + a$].

---

[9]See Grassmann (1861), Peano (1889), Poincaré (1902).

[10]Nelson's argument is a bit opaque. The translation above was suggested by one of the two anonymous referees appointed by Springer. With much appreciation I put it here instead of my original one.

The problem is: how many apples are in the left-hand basket to be counted after following instruction (11)? Well, what equation (A), taken as a definition, says is that there are in the left-hand basket as many apples as we would obtain if there had been a apples in the left-hand basket and only one in the right-hand basket and if we followed instruction (12):

(12) You first add b apples to the left-hand basket $[b + a]$. Then you put the one apple in the right-hand basket into the one on the left $[1 + (b + a)]$.

The principle (A) says that the number of apples in the left-hand basket in operation (12) is the same as we would obtain in operation (11). It is clear that there is no way we can make this work by defining words.

References

Collingwood, Robin G. 1939. *An autobiography.* Oxford: Clarendon Press.
Frege, Gottlob. 1884. *Die Grundlagen der Arithmetik: eine logisch-arithmetische Untersuchung über den Begriff der Zahl.* Wrocław: Wilhelm Koebner. [English translation by J.L. Austin: *The foundations of arithmetic: a logico-mathematical enquiry into the concept of number.* Oxford: Blackwell (1950). New translation by Dale Jacquette: *The foundations of arithmetic.* New York: Pearson, 2007].
Grassmann, Hermann. 1861. *Lehrbuch der Arithmetik für höhere Lehranstalten* [*Handbook of arithmetic for institutions of higher education*]. Berlin: Enslin.
Heilborn, Paul. 1912. *Grundbegriffe des Völkerrechts* [*Fundamental concepts of international law*]. Stuttgart: Kohlhammer.
Kaufmann, Erich. 1911. *Das Wesen des Völkerrechts und die clausula rebus sic stantibus: rechtsphilosophische Studie zum Rechts-, Staats- und Vertragsbegriff* [*The essence of international law and the clausula rebus sic stantibus: A philosophical study of the concepts of law, state, and contract*]. Tübingen: Mohr.
Keynes, John M. 1919. *The economic consequences of the peace.* London: Macmillan.
Leibniz, Gottfried Wilhelm. 1704. *Nouveaux essais sur l'entendement humain.* Manuscript published posthumously. [English translation: *New essays on human understanding.* Cambridge: University Press (1996)].
Peano, Giuseppe. 1889. *Arithmetices principia nova methodo exposita.* Torino: Bocca. [English translation: The Principles of Arithmetic, Presented by a New Method, in Jean van Heijenoort, *From Frege to Gödel: A source book in mathematical logic, 1879–1931*, 85–97. Cambridge (MA): Harvard University Press (1967)].
Poincaré, Henri. 1902. *La science et l'hypothèse.* Paris: Flammarion. [English translation: *The foundations of science: science and hypothesis, the value of science, science and method*, 27–197. New York: The Science Press (1929)].
Schoen, Paul. 1915. Zur Lehre von den Grundlagen des Völkerrechts [Contribution to the theory of the foundations of international law]. *Archiv für Rechts- und Wirtschaftsphilosophie* 8: 287–321.
Zitelmann, Ernst. 1914. Haben wir noch ein Völkerrecht? [Do we still have an international law?]. *Preussische Jahrbücher* 158: 472–495.

Lecture XVIII

Abstract Some famous authors in the philosophy of physics (e.g. Dingler) have also incurred in the concept-swapping fallacy, witness his attempt to prove the law of causality (a synthetic judgment) by means of definitions. But it is in the so-called 'theory of knowledge' or 'epistemology' that we can see the fallacy at work from the very start. Although 'epistemologists' claim to be Kant's successors, their very field of inquiry is incompatible with the critique of reason, and in fact is an egregious case of trying to derive synthetic judgments from mere concepts. The problem they set out to solve, viz the validity of our principles, requires an utterly different kind of inquiry, the one Kant called 'deduction'.

My last lecture ended with an example taken from arithmetic. I would like to add another example taken from the philosophical foundations of physics (from natural philosophy, as we could also say). I take it from Hugo Dingler's book *The Foundations of Applied Geometry* (1911). The author is an applied mathematician from Munich who knows that the usual empiricism in physics is untenable and is looking for an alternative. His approach is essentially Poincaré's conventionalism, but I have chosen this example not because it is typical of the conventionalist school of thought but rather because it illustrates the fallacy we have been discussing of using a nominal definition in order to prove synthetic propositions. The proposition whose proof we want to inspect belongs to the very foundations of natural science: it is the law of causality. Because of the manner in which he introduces it, Dingler calls it the 'principle of identity' or the 'law of identity'. He calls it that because he presents it as a direct application of the law of identity in logic. In order to carry out the proof Dingler makes use of the following definitions[1]:

---

[1] See Dingler (1911, Chap. II, §5, p. 51, Definitions 6 and 7). Nelson's quotation (probably from memory) is not quite faithful to the original. Given that his argument is not affected by it, the text has been corrected and variables have been introduced to clarify the definition.

(1) Any part of a process P is called *a circumstance of P*.
(2) Any circumstance c [of a process P] is called *essential to*, or *a condition of*, P if [and only if] whenever c changes P also changes. Otherwise c is called an *unessential* circumstance of P.

Before I come to the point, it is instructive to note that these definitions already contain a contradiction. The definitions are used to define the task of physics as finding out the essential circumstances of a process by means of experimentation. Yet according to these definitions there can be no such thing as an unessential circumstance of any process. For if every circumstance is part of a process, then no circumstance can change without changing the total process. Every circumstance of a process is therefore essential to it or a condition of it. The task of establishing the essential circumstances of natural processes, i.e. of distinguishing them from their unessential circumstances, contains an inherent contradiction. But this is not the point I want to make here and I mention it only in passing. My point is that it is dangerous to try to establish concepts via definition when those concepts are already given before any definition. All the concepts mentioned (CIRCUMSTANCE, ESSENTIAL, CONDITION, PROCESS) are already given to us, we possess them all before any definition, and hence there are original meanings for all of them; and it is quite true that we distinguish between essential and unessential circumstances in natural processes.

Let us now consider the practical application of Dingler's definition. From this definition he immediately argues[2] the *consequence* that

Under the same conditions the same process occurs.

This is his so-called law of identity.[3] It means that we have only to reproduce the essential circumstances of a process in order to make sure that the process itself is repeated.

According to Dingler, the whole of physics is possible because of this law. The conditions are also called 'the cause' and the process itself 'the effect', which is the way we speak in ordinary language. The consequence Dingler draws from his definitions certainly follows from it, for they say that the effect itself belongs to the essential circumstances of a process, and so they postulate that, if the same conditions are produced, then the effect will be reproduced. Predicting that effect is thus to say the same thing all over again. A PROCESS was defined as identical with THE TOTALITY OF ITS ESSENTIAL CIRCUMSTANCES. The consequence of this is that in order to know whether all the conditions of the process are present in full, we have to know already that the effect occurs. But if we already know that the effect occurs, then by virtue of the *logical* law of identity we can repeat that the effect must occur. We do not need Dingler's proof to know that. On the basis of Dingler's 'law of identity' we cannot say that the effect will occur without presupposing that it always does.

---

[2] See Dingler (1911, 52).
[3] See Dingler (1911, 53, 54).

Lecture XVIII

Let us put Dingler's 'law of identity' into ordinary language by setting aside his misuse of a contrived nominal definition. What it says is simply (3):

(3) If not only the other so-called conditions of a process but also the effect itself takes place, then the effect takes place.

That is the exact, precise meaning of Dingler's 'law', based on the two definitions that precede it. We can say this more briefly and in a less roundabout way as in (4):

(4) We only need to repeat a process in order to make sure it is repeated.

This 'law' can of course never serve the purpose for which it was introduced here, namely to predict the effect of any natural process without having to rely on experience.

This will suffice as a collection of examples for the said fallacy. Nevertheless, I do not want to close my study of it quite yet, for it continues to play a significant role in a somewhat disguised form—as part of a philosophical discipline that goes by the name of 'epistemology' or 'the theory of knowledge'.

It is well known that this discipline can historically be traced back to Kant's *Critique of Reason*. People no longer use the phrase 'critique of reason' but talk instead of 'the theory of knowledge' or of 'epistemology', because they correctly feel that the kind of questions being asked has changed. We have to thank Kant for unravelling the mystery that underlies the misuse of nominal definitions in philosophy and thus for giving an explanation for the whole illusion of dogmatic metaphysics. However, given that Kant brought to light the underlying fallacy leading to this illusion in his *Critique of Reason*, it is amazing that a discipline that sees him as its founder is based entirely on repeating that fallacy.[4] For I believe this is what everyone does who nowadays calls himself an 'epistemologist'. Epistemologists work on the question of the validity of our knowledge, and indeed not of this or that single instance of knowledge, but rather of knowledge in general; and they see themselves as the successors of the founder of the Critique of Reason precisely because they work on this problem. Of course, not everyone defines the concept of EPISTEMOLOGY in this precise way. It may be the case that this or that author uses the word to denote a different endeavour. Yet on the whole the problem of the validity of any and all knowledge dominates the field called 'epistemology'.

To that I say that all possible attempts to solve this problem depend on repeating in new forms the same fallacy that Kant brought to light, namely a surreptitious swapping of concepts allowing one to prove a synthetic proposition without using another one as a principle. What epistemologists want to do is to solve the problem of the validity of any and all knowledge. To do this they need a criterion of epistemic validity. But there cannot be any such criterion. The very concept of CRITERION OF EPISTEMIC VALIDITY is a contradiction in terms. For we have two choices: *Either such a criterion is itself an instance of knowledge or it is not.* If it were an instance of knowledge, then it would belong to the problematic class of those things

---

[4]For a historical study of how the Kantian 'theory of knowledge' developed, see Nelson (1912).

whose validity would first have to be established. Hence the criterion cannot itself be an instance of knowledge. If it were *not* an instance of knowledge, then it would be external to knowledge and we would have to know it in order to apply it as a criterion. But for this knowledge of the criterion to be considered valid we should have applied the criterion to it beforehand. In other words, it should already be part of what we know and have already been validated as knowledge—in contradiction to the very concept of such a criterion. And so there cannot be any epistemological criterion, and the task of finding one contains a contradiction.

To illustrate this argument, assume that the pragmatic criterion of truth proposed by conventionalism is our epistemological criterion. As such it is not an instance of knowledge and can be used to check the validity of any and all knowledge. The criterion says that convenience decides whether anything is true. To introduce it as an epistemological criterion it would be have to be known as such. Yet to know it as such is to know that it is convenient, for only thus we would know it as true. But to know it as convenient we need to assume the criterion of validity, thus of the proposition which we are trying to prove.

To illustrate the opposite case, when the epistemological criterion is itself an instance of knowledge, we can use the epistemology which proposes self-evidence as such a criterion. The clearest exposition of this kind of epistemology appears in the book *On the Empirical Foundations of our Knowledge* (1906) by the Austrian philosopher, Alexius von Meinong. He says: in order to show us that self-evidence is a criterion of the type we are seeking, it would have to be self-evident that it is true that self-evidence is a criterion of truth. And indeed Meinong says that this proposition *is* self-evident. Therefore self-evidence is a criterion of truth. Examples like this one can be piled up so as to overwhelm anyone. I could fill up several terms worth of lectures with lists of such examples. Yet we do not have so much time and so I content myself with referring those who are interested to my book *On the So-Called Problem of Knowledge*.[5] I prefer to talk in this course about things that cannot be learnt from books.

It can be quite easily shown that the entire endeavour of 'epistemology' is actually no more than a repetition of logicistic metaphysics. To solve the problem it raises, the theory of knowledge cannot possibly assume any instance of knowledge as given. All candidates to the status of knowledge are suspect. So its starting-points cannot be positive assertions but must be problematic representations, i.e. mere concepts; and should the epistemological problem be amenable to a solution, that solution would amount to deriving synthetic propositions from mere concepts; and so, since only analytic judgments can emerge from mere concepts, the solution would be a matter of converting analytic judgments into synthetic ones. For it is after all clear that the assertion 'We possess knowledge' or 'We have valid knowledge' can only be a synthetic judgment. And besides, the purpose of epistemology is to show the validity of our knowledge and so the possibility of furthering knowledge beyond mere concepts.

---

[5] See Nelson (1908; Meinong is examined and found wanting in Chap. IV, §§21–24).

Lecture XVIII

The importance of the present fallacy in the context of 'epistemology' can also be made plain from another perspective. The endeavour to find *reasons* for all knowledge ultimately entails that there can be no such thing as immediate [or direct] knowledge. For the postulate that all knowledge must have a reason can only mean that any instance of knowledge has to be reduced to another instance of knowledge and so on in such a way that no piece of knowledge is firm if it does not rest on some other piece of knowledge. It is thus tacitly presupposed that there is no such thing as immediate, i.e. unsupported knowledge. Every piece of knowledge is assumed to be mediated. Mediated knowledge is judgment, and if one can support a judgment only by another judgment (i.e. by another piece of mediated knowledge), then to support a judgment would always mean to *prove* it [for proof is reduction of one judgment to another judgment]. It thus becomes clear that 'epistemology' is the attempt to prove all knowledge, and we have already seen that this task contains a contradiction.

But how can we explain that, if Kant conclusively refuted the errors of logicistic metaphysics, people believe that the repetition of those errors in 'epistemology' has its origin precisely in Kant? Here is why. In the *Critique of Reason*, a method (called 'critical') had been put forward as the appropriate method for philosophy. That method is *regressive* in the sense that we go back from the particular to the general, and *analytic* in the sense that we take apart any given judgment so as to find its presuppositions. In this manner we arrive at certain highest and most general presuppositions of those particular judgments with which our analysis started. The method, however, only shows that certain judgments are presupposed by certain other judgments. There is absolutely no way we can say that the judgments we start with are the foundation or the reason of the judgments we show to be presupposed by the former. It is the other way round, the judgments we start with rest on their presuppositions, for it is *from* those presuppositions or principles that we prove the judgments. To pretend otherwise would be circular. Yet those highest and most general presuppositions of our judgments are themselves judgments and so also in need of support. That is precisely what the postulate says: to hold a judgment true, it has to have a reason, and we have to show that it does have one. Every judgment must be reduced to a piece of knowledge that contains its reason.

Fine, but if we formulate it like that, we are only saying that all *judgments*—all mediated instances of knowledge—need support.[6] We are in no way saying that all knowledge needs support. The logical principle of sufficient reason requires only that all mediated knowledge must in the end be reduced to immediate knowledge. To hold a judgment true, we have to show that it recapitulates a piece of immediate knowledge. The postulate applies also to those highest and most general judgments that the critical method discovers by the logical analysis of the particular judgments that serve as our starting points. They are the most general presuppositions, the last

---

[6]I introduce here a slight amendment in the German text, which has two sentences that opaquely anticipate the reasoning that follows. Nelson comes back to this argument in his last lecture. The argument was discussed in depth by Popper (1979).

grounds that prove all particular *judgments*, yet not the last grounds of all our *knowledge*. They are themselves judgments and so we must give reasons to support them, otherwise we would be rightly attacked as dogmatic, even though we could at least plead the extenuating circumstance that we did not put forward those highest judgments in an arbitrary fashion, but by applying the critical method we came to them via the analysis of particular judgments. Nonetheless, as I said before, they have been shown to be factual presuppositions of particular judgments and thus far unsupported. They should thus be reduced to the underlying immediate knowledge.

A huge problem for logicians emerges here. What kind of support or backing up or giving of reasons should this be? A proof perhaps? That would mean reducing the highest judgments to logically higher ones—a contradiction in terms, for HIGHEST JUDGMENTS are defined as JUDGMENTS THAT CANNOT BE LOGICALLY REDUCED TO HIGHER ONES. Their being self-evident perhaps? They are certainly not *logically* self-evident, so the knowledge underlying them would have to be as immediate and clear as sensory intuition. Kant used the term 'demonstration' for the reduction to sensory intuition, so it seems that backing up the highest judgments would be demonstrating them in this Kantian sense.[7] Now this is very tempting, for something like that seems to be the difference between judgments as mediated pieces of knowledge and any non-judgmental kind of knowledge. The mediated character of judgment is a matter of using concepts. In judgments we come to know objects by means of concepts, by classifying them under general concepts. A piece of immediate knowledge underlying a judgment seems to be non-conceptual and thus intuitive. But I have repeatedly shown before that philosophical judgments are not demonstrable in Kant's sense, for they are not immediately self-evident and we have no intuition of them. The dogmatic fallacy consisted precisely in ignoring this. For traditional logic, however, to give a reason for something is either to prove it or to demonstrate it [in Kant's sense of 'demonstration'].

Kant saw the deep problem here and set about the task of subjecting the philosophical principles (after having uncovered them as principles via the regressive method) to a 'deduction', as he called it. This term of Kant's is all too easily misunderstood, the more so today where it tends to be used to denote a kind of proof. But the support we are looking for cannot be a proof. So, if proof and demonstration really are the only possible means of backing up a judgment, then we must conclude that the philosophical principles, being neither provable nor demonstrable, have no means of support. This would mean that in philosophy we can never get beyond a dogmatic assertion of its principles, so that doing philosophy is in the end an arbitrary affair. We might at best make use of conventionalist and similar perspectives in order to show the convenience for this or that purpose of making this or that arbitrary assumption, but there would not be any foundational task to be fulfilled.

---

[7] I keep the word 'demonstration' against English usage because it is introduced as Kant's expression. A word like 'ostension' would perhaps be more appropriate and better understood. The Kantian distinction is similar, albeit not identical, to Russell's distinction between 'knowledge by acquaintance' and 'knowledge by description' (see Russell 1911).

References

Dingler, Hugo. 1911. *Die Grundlagen der angewandten Geometrie: eine Untersuchung über den Zusammenhang zwischen Theorie und Erfahrung in den exakten Wissenschaften* [*Foundations of applied geometry: An inquiry into the connection between theory and experience in exact science*]. Leipzig: Akademische Verlagsgesellschaft.

Nelson, Leonard. 1908. Über das sogenannte Erkenntnisproblem [On the so-called problem of knowledge]. *Abhandlungen der Fries'schen Schule (N.F.)* 2(4): 413–818 [Reprinted in Nelson (1971–1977), vol. II, pp. 59–393].

Nelson, Leonard. 1912. Untersuchung über die Entwicklungsgeschichte der Kantischen Erkenntnistheorie [Historical enquiry into the development of the Kantian theory of knowledge]. *Abhandlungen der Fries'schen Schule (N.F.)* 3(1): 33–96 [Reprinted in Nelson (1971–1977), vol. II, pp. 405–457].

Nelson, Leonard. 1971–1977. *Gesammelte Schriften*, 9 vols. eds. Paul Bernays, Willy Eichler, Arnold Gysin, Gustav Heckmann, Grete Henry-Hermann, Fritz von Hippel, Stephan Körner, Werner Kroebel, and Gerhard Weisser. Hamburg: Felix Meiner.

Popper, Karl. 1979. *Die beiden Grundprobleme der Erkenntnistheorie*. Tübingen: Mohr [English Translation: *The two fundamental problems of the theory of knowledge*, London, Routledge, 2008. The original German text was written by Popper in the late 1920s or early 1930s].

Russell, Bertrand. 1911. Knowledge by acquaintance and knowledge by description. *Proceedings of the Aristotelian Society (N.S.)* 11: 108–128.

von Meinong, Alexius. 1906. Über die Erfahrungsgrundlagen unseres Wissens [On the empirical foundations of our knowledge]. *Abhandlungen zur Didaktik und Philosophie der Naturwissenschaften* 1(6): 1–113.

Lecture XIX

Abstract Kant's Critique of Reason was a new field purporting to inquire into a priori knowledge. His self-appointed successors invented another field, 'epistemology', which they believed to be itself a priori. To convince themselves of this they followed a principle (first made explicit by Kuno Fischer) that a priori knowledge cannot be known a posteriori. This principle is a synthetic judgment which was widely accepted because it was disguised as an analytic judgment. It is interesting that by the same fallacy one can arrive at the opposite conclusion ('empiricist epistemology'). And in legal philosophy we can see that the dispute about the League of Nations has the very same structure.

The ambition of my last lecture was to explain to you to what extent the philosophical discipline known as 'epistemology' or 'the theory of knowledge' is just a new form of logicistic metaphysics or the attempt to derive synthetic judgment from mere concepts. We can also describe the topic of epistemology by saying that it inquires into the relation between knowing and the known. For if we ask whether knowledge is valid, objectively valid, then we are not asking about the relation between two pieces of knowledge but rather about the relationship between the whole of our knowledge and all objects. If we show that one piece of knowledge agrees with another, where one is immediate and the other mediated, we have proved absolutely nothing about the objective validity of knowledge. For despite the inner agreement of all pieces of knowledge with each other, there need not be an agreement of knowledge with objects; and despite all inquiries into the nature of the internal relations between the different things we know and how far they agree among themselves, nothing follows as to the relation between knowledge as such and the objects themselves. For that is a question about the objective validity of our knowledge. All giving of reasons to back judgments by means of either proof or demonstration has merit only to the extent that the item of knowledge [another judgment in the case of proof, a sensory intuition in the case of demonstration] we are using to support the judgment is itself objectively valid, i.e. to the extent that it agrees with its object. It is this presupposition that epistemology makes into a problem of inquiry.

The question arose first in connection with philosophical principles, and it is easy to see why. For both proof and demonstration fail here as means of support.

I have not shown in detail that philosophical principles can never be proved. It is perfectly clear that they cannot be proved within their own field of knowledge. But the question arises whether a metaphysical principle—i.e. a synthetic philosophical principle—may not perhaps be reduced to a theorem of some other science. We can easily convince ourselves that this is also impossible, for what kind of theorem could that be? If it is analytic, we know that no synthetic proposition can be proved by it; and if it is synthetic, then there are again two options. It is either empirical, but then no philosophical conclusion can be derived from empirical premises, for the former are independent of all experience; or else it is itself a metaphysical proposition and then the proof would be a logical circle. There is no other option. On the other hand, no demonstration of the metaphysical principles is possible, because there is no intuition that could support them. If there is any metaphysical truth, then it can only become clear to us by means of thinking and in the form of a judgment. So philosophical principles are not demonstrable either.

The law of causality, which we discussed earlier, helps illustrate the argument. The law of causality is plainly a metaphysical judgment that underlies the possibility of all empirical inferences [about what causes what], hence cannot itself be derived from experience. It also is doubtless a synthetic proposition, so that any attempt at proving it in the logicistic manner discussed above would inevitably be in vain. Both means of finding support for the law of causality—proof as well as demonstration—fail utterly. There is thus no way of supporting, backing or founding a metaphysical principle for judgments, not even by reducing it to some underlying immediate knowledge, and here arises the problem of their relation to the object. This is how we pass over from the problem of the Kantian *Critique of Reason* to the problem of modern epistemology.

I will now describe a notorious argument to illustrate the fact that every single time epistemologists manage to avoid moving in circles, it is by means of the logicistic fallacy we have earlier identified in a different form and used for different purposes. Those who have become famous in the history of philosophy as the successors of Kant re-worked the critique of reason into a theory of knowledge, in other words they developed a supposedly scientific study of the relation of knowledge to its object on a rationalistic foundation. They not only considered philosophy itself—the system of philosophy, in particular the system of metaphysics—to be a *rational* science, but the critique of reason, or rather the so-called theory of knowledge, was to be considered a *rational* science as well.[1] In other words, that knowledge by means of which we would give reasons for the

---

[1] Two terminological remarks may be useful. On the one hand, philosophy as a system contains for Nelson two subsystems, viz. logic and metaphysics. Logic consists only of analytic propositions and metaphysics only of synthetic a priori propositions. Metaphysics in its turn consists of two parts, the first concerning the natural world and the second ethical questions (the 'realm of freedom'). On the other hand, the adjective 'rational' refers in this context to the distinction between 'rational' and 'empirical' as usual in the Leibniz-Wolff school. We should distinguish this sense from the related but different and later usage by which people talk about 'rational mechanics' in the sense of a mathematical deductive science (see e.g. Truesdell 1968, 1991).

Lecture XIX

metaphysical principle should itself have a metaphysical character. The absurdity of such an endeavour becomes apparent as soon as it is so clearly expressed. For the question of whether metaphysics is at all possible must of course be decided elsewhere, not in the frame of a new-fangled form of metaphysics.

Some authors identified the fallacy and suggested that one could do epistemology from an empiricist perspective. I say 'empiricist' and not just 'empirical', for the idea was that not only did the theory of knowledge have to be an empirical science but the whole of philosophy should also be part of empirical knowledge (a strange kind of philosophy, by the way). According to this school of thought, philosophy is entirely derived from epistemology, not from some [inexistent] immediate knowledge. Philosophy could thus become a science, and an empirical one to boot, but only as a derivation from the empirical theory of knowledge. The question arises as to the relation between the way of knowing of epistemology (or the critique of reason) and the way of knowing of philosophy as a system [of propositions] backed up by epistemology.

This question first became famous through a talk given by Kuno Fischer, the famous historian of nineteenth-century German philosophy, on the occasion of his becoming vice-chancellor of the University of Jena. In that talk, called 'The Two Kantian Schools in Jena' (1862), Kuno Fischer asks what kind of knowledge is then the Critique of Reason, and argues against those who hold it to be empirical in character. He ends by taking the side of the rationalist school—the kind of metaphysics elaborated by Fichte, Schelling and Hegel, as well as their successors. What for Fischer clinches the argument is the celebrated dictum[2]:

> [Fischer's dictum:] What is *a priori*, can never be known *a posteriori*.

Metaphysical knowledge is to rest on the critique of reason, but because it is a priori knowledge, it cannot be known a posteriori, for that would be, or so argues Kuno Fischer, a contradiction in terms. If the critique of reason were empirical, this would contradict the a priori character of the metaphysical knowledge that should rest on it. Fischer thus only needs the principle of non-contradiction to prove his point. The assumption that the critique of reason is an empirical way of knowing is shown to contain a contradiction. The contrary assumption [that the critique of reason is a priori] is also a matter of pure logic. Fischer's argument seems to run thus:

(1) Metaphysical knowledge is a priori. (Remember the example of our knowledge of the causal law.)
(2) What is a priori cannot be known a posteriori.
(3) Therefore, the critique of reason (as knowledge of the a priori) is itself a priori.

I hope all of you are able to point out the fallacy in this argumentation. The weak point lies in the phrase 'what is a priori' in (2). Well, what is it? What is a priori? Kuno Fischer does not provide his own explanation; he refers to Kant's own

---
[2]This celebrated and oft-repeated dictum can be found on p. 99 of Fischer (1862).

linguistic use. Accordingly, this expression relates to the modality of knowledge. We would therefore have to reformulate Fischer's dictum in a more precise way as follows:

(4) A priori knowledge (for only knowledge can have such an attribute) cannot be known a posteriori.

That would be the meaning of the dictum [proposition (2) above] It seems self-evident because the concept A POSTERIORI is defined by negation—a priori and a posteriori are mutually exclusive attributes. Instead of 'a posteriori' I could write 'not a priori'. Thus it is clear that a priori knowledge cannot be a posteriori knowledge. Otherwise we would have a contradiction. Of course we admit that:

(5) A priori knowledge is not a posteriori knowledge.

But from (5) [which is a definition] does (6) follow?

(6) A priori knowledge cannot be known a posteriori.

Not in the least, since the vagueness of the expression 'what is a priori' in (2) allows one concept—A PRIORI KNOWLEDGE—to be swapped for a very different one—KNOWLEDGE OF A PRIORI KNOWLEDGE, i.e. knowledge which has a priori knowledge as its object. Remember how the very concept of a CRITIQUE OF REASON is defined:

CRITIQUE OF REASON = KNOWLEDGE WHOSE OBJECT [OF INQUIRY] IS A *PRIORI* KNOWLEDGE.

The moment one looks at this definition, the whole argument of Kuno Fischer crumbles away. From the premise that a priori knowledge cannot be a posteriori knowledge it does not follow that some other knowledge could be a posteriori, namely that knowledge by means of which we grasp that this or that item of knowledge is a priori. It is this fallacy which sustains the illusory renewal of rationalistic metaphysics in the hands of Kant's [self-appointed] epistemological successors.

The opposite argument rests on the same fallacy. The empiricist school has tried to renew dogmatic metaphysics by assuming also that a priori knowledge cannot be grasped a posteriori. It is the very same prejudice that rationalism and empiricism share. Consider the logically equivalent contraposition of Fischer's dictum:

(7) What is known a posteriori, cannot be a priori knowledge, i.e. the object of a posteriori knowledge cannot be a priori knowledge.

If there are two items of knowledge such that item *A* is about item *B*, and item *A* is a posteriori, then [so the argument goes] item *B* cannot be a priori. All empiricist epistemology rests on this premise. It commits the same swapping of concepts that rationalists are guilty of: Since an empiricist critique of reason has the task of reducing the system of philosophy to its underlying reasons, then the a posteriori character of the former is transferred to the latter. This system of philosophy can then contain no a priori knowledge, and so philosophy becomes itself an empirical science.

There is a deeper reason for this dilemma [rationalism vs. empiricism] in which post-Kantian philosophy in the form of epistemology got entangled. It did not arise

Lecture XIX 171

by chance. It is rooted in the very way in which epistemology poses the question. I have already shown that epistemology can lead only to a renewal of logicistic metaphysics; but this means that it consists in an attempt to *prove* the metaphysical principles, to reduce them to higher premises, namely to epistemological theorems. And if that were the case, then it would of course be true that the epistemic nature of the consequent [being either a priori or a posteriori] would follow from the epistemic nature of the antecedent, and the other way round. To prove a piece of a priori knowledge we need a piece of a priori knowledge, or else we would have a contradiction; and if our premises are items of a posteriori knowledge, then whatever we prove from them will be an item of a posteriori knowledge.

There is no escape from this dilemma. Every attempt to avoid it through compromise would throw us into a contradiction. The premises that allow us to prove a proposition must be of the same epistemic kind as the proposition proved. If the one belongs to a posteriori knowledge, the other cannot belong to a priori knowledge, or the other way round. Yet this dilemma depends solely on assuming that the relation between epistemology and epistemology-supported metaphysics is the logical relation that exists between premises and conclusion; an assumption forced by traditional logic, according to which, apart from demonstration (which is indeed out of the question), the only way to back a judgment is by means of proof. But we can get out of the dilemma by relinquishing this prejudice.

Whether and how this is possible I will leave aside for now.[3] All I wanted was to show you by example that epistemology can only achieve the illusion of a successful solution to its problem by following in the footsteps of the much older logicist metaphysics, i.e. by abusing nominal definitions, by swapping analytic judgments for synthetic ones. Proposition (5) in Fischer's argument above is analytic, a nominal definition of the concept A PRIORI. Proposition (6) is synthetic and can never be reduced to (5). In fact, (6) is false. I have not yet shown that (6) is false, but it is easy to see. To possess a priori knowledge is undoubtedly a *fact* that can only be established as such, hence a posteriori. It is factual, not necessary knowledge. From a mere concept of A PRIORI KNOWLEDGE nobody would dare to prove to possess such knowledge. So a paradox emerges—the critique of reason is empirical and the metaphysics supported by it is a priori.

This example illuminates the dilemma of the post-Kantian school of epistemology and reminds us of our first example concerning geometry (Chapter "Lecture VIII", Fig. 2). We can use it again for a different purpose. It is a typical consequence of misusing nominal definitions in philosophy that to any pseudo-proof obtained by such misuse we can oppose an equally good pseudo-proof of the contrary. Both pseudo-proofs share a common hidden presupposition which, equally concealed from both parties, gives birth to an antinomy, an irresolvable conflict. Once the presupposition is in place, the antinomy is unavoidable. The logical necessity

---

[3]Nelson's way out of the dilemma was the subject of his dissertation (1904) and the starting point, never relinquished, of all his philosophical efforts. Because Nelson's way out was pretty controversial among his contemporaries, he prefers in these methodological lectures to avoid going into it.

claimed by one of the two opponents is the same logical necessity claimed by the other—which again explains why discussing whether or not the other party's argument is valid does not lead anywhere. If one party takes a true analytic premise and manages to use it to argue for a false synthetic conclusion, then by the same logic the other party can take the falseness of such a conclusion to argue that the premise is false. The first party's argument starts by correctly attributing a predicate to the objects falling under a concept C, but then it illegitimately replaces C with a different concept D and concludes with a false attribution of the said predicate to the objects that fall under D. The second party inverts the argument and, having [correctly] denied that the predicate applies to the objects that fall under D proceeds to deny it also from the objects that fall under C. To make clearer how this works we can again use a diagram. Let us now use the example discussed in my last lecture, as in Fig. 1.[4]

You have here the assumption that particular governments have the choice either to live in anarchy with each other or to sacrifice themselves as states in order to merge into a world state. In other words, the alternative of inter-state anarchy or world government is assumed to exhaust all possibilities, to be [what logicians call] a 'complete disjunction'.[5] Such an assumption is unconscious. Neither opponent is aware of it.

One of the parties starts from the unacceptability of anarchy (the self-evident illegitimacy of a situation of anarchy holding between particular governments) and concludes that a world government is necessary, that we are bound by justice to create one. Particular governments are expected to sacrifice themselves and to vanish in order to avoid anarchy.

On the other side of the argument it is recognised that this demand [self-sacrifice of particular governments] is in no way justified. The necessity of a world government is rejected with good reason (as good as one had to reject inter-state anarchy) and it is concluded from this (and the argument is at least as valid) that anarchy between governments should be preserved, or more clearly, and in the negative (as in Fig. 1), that inter-state anarchy is not unacceptable. Such are the pronouncements of the advocates of an international rule of force, as I called them when I gave you some quotations in my last lecture.

As soon as we let fall the common assumption, we recognise that both arguments fail—that in spite of appearances neither is logically valid. Only thanks to this shared and unconscious assumption do they arrive at their conflicting results.

---

[4]The reader will notice that five of the six boxes in Fig. 1 contain noun phrases instead of full sentences (see Footnote 6 in Chapter "Lecture VIII"). This is immaterial, for one could replace the former with the latter at any time. The diagram contains thus three inferences, which can be read as follows: (1) The complete disjunction between inter-state anarchy and the world-state together with the unacceptability of inter-state anarchy implies the necessity of a world-state; (2) The complete disjunction between inter-state anarchy and the world-state together with the denial that a world-state is necessary implies the denial that inter-state anarchy is unacceptable; (3) The unacceptability of inter-state anarchy together with the denial that a world-state is necessary implies that the disjunction between inter-state anarchy and the world-state is incomplete.

[5]See Footnote 3 in Chapter "Lecture VIII".

Lecture XIX

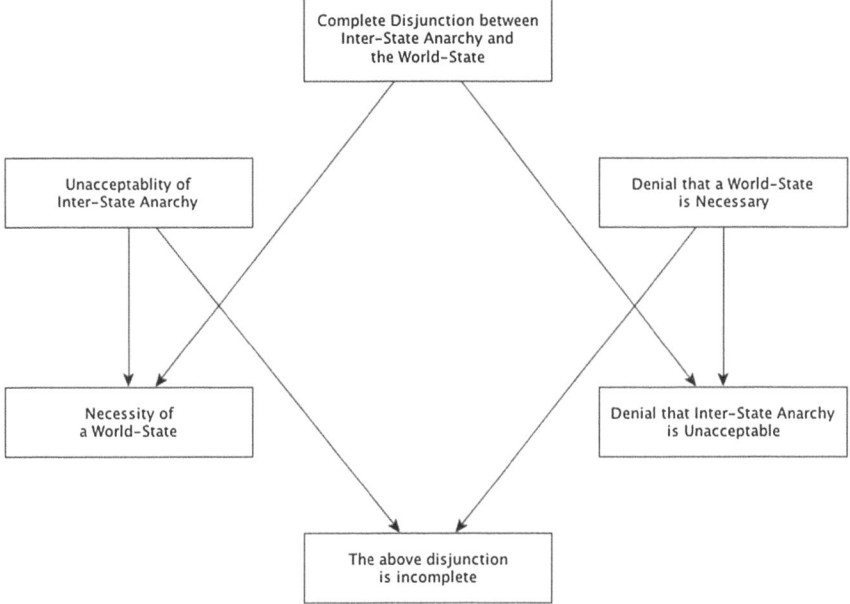

Fig. 1 A false dilemma in political philosophy

One need only grasp the point once to bring about immediately the otherwise seemingly impossible or hopeless reconciliation of the two parties. One need only let go of the assumption that I placed on top in Fig. 1 and that has no justification in the world—one need only see once that the disjunction is incomplete—in order also to see that states can continue to exist as states after giving up their sovereignty, just as individuals keep their rights even though these rights are protected by the police, the laws and the courts.

The conclusions drawn by our two opponents depend solely on the assumed completeness of that disjunction; and that disjunction *seems* to be not only true but logically necessary because of the surreptitious swapping of concepts brought about by the arbitrary nominal definition we identified in our last lecture, viz the arbitrary definition of the concept of STATE or the concept of INTERNATIONAL LAW, by which the predicate that belongs to the objects subsumed under the concept of SOVEREIGN, is surreptitiously attributed to the objects of the original concept of STATE.

We can always treat the hidden assumption as directly asserting an equivalence between the two concepts [viz STATE and SOVEREIGN].[6] In other words, the assumption would actually assert that propositions (8) and (9) are incompatible:

---

[6]The following argument is stated by Nelson in an excessively abstract form. To facilitate understanding I have throughout replaced the abstract terms with the concrete example that Nelson has in mind.

(8) Anarchy between *states* is unacceptable.
(9) A world *state* is not necessary.

Now, propositions (8) and (9) would only be incompatible if STATE = SOVEREIGN. Yet if these two concepts are not identical but only appear to be, then there would not be a contradiction between (8) and (9) at all.[7] Affirming A contradicts denying A, but it does not contradict denying B. However, if it is asserted that A and B are the same, viz STATE and BEING SOVEREIGN, then affirming it of an object [existing states] is incompatible with denying it of the other object [the world state]. But if the identity [as I hold] is only apparent, then the contradiction goes away; for it can only be sustained as long as we have not yet discovered the difference between the two concepts. All we need is to show the illusion of that equivalence once, and we are ready, in one fell swoop, to get rid of all the irresolvable antinomies that have divided philosophers since time immemorial.

References

Fischer, Kuno. 1862. *Akademische reden* [Academic Addresses]. Stuttgart: Cotta.
Nelson, Leonard. 1904. Die kritische Methode und das Verhältnis der Psychologie zur Philosophie: Ein Kapitel aus der Methodenlehre. *Abhandlungen der Fries'schen Schule (N.F.)* 1(1): 1–88. [Reprinted in Nelson (1971–1977), vol. I, pp. 9–78. English translation: The Critical Method and the Relation of Philosophy to Psychology: An Essay in Methodology, in *Socratic Method and Critical Philosophy: Selected Essays* (pp. 105–157), New Haven, Yale University Press, 1949].
Nelson, Leonard. 1971–1977. *Gesammelte schriften*, 9 vols. eds. Paul Bernays, Willy Eichler, Arnold Gysin, Gustav Heckmann, Grete Henry-Hermann, Fritz von Hippel, Stephan Körner, Werner Kroebel, and Gerhard Weisser. Hamburg: Felix Meiner.
Truesdell, Clifford. 1968. *Essays in the history of mechanics*. New York: Springer.
Truesdell, Clifford. 1991. *A first course in rational continuum mechanics*, vol. I, 2nd ed. Boston: Academic Press.

---

[7]Indeed, if anarchy between *sovereign* states is unacceptable—here Nelson would agree with the pacifists—then a world *sovereign* is necessary—here Nelson parts company with the pacifists. And again, if a world *sovereign* is *not* necessary—here Nelson would agree with the legal scholars he criticises—then anarchy between *sovereign* states would be perfectly acceptable—here Nelson parts company with the legal scholars.

Lecture XX

Abstract All the analysed examples show how typical the philosophical fallacy is that consists in replacing a given concept by a made-up one with the result that we seem to have proved a synthetic judgment on the strength of a purely analytic one, in fact a mere nominal definition. Kant was the first master in revealing the fallacy in his doctrine of the cosmological antinomies, which rely on falsely identifying our concept of nature and the transcendental idea of universe. Now, all the analysed examples belong to philosophers who construct arguments, even if fallacious ones. What about those who instead of arguing just give us their intuitive wisdom? It can be shown that their illusory systems also depend on the concept-swapping fallacy, albeit in a more insidious form.

I hope that the series of examples that I have put forward, developed, and explained have made clear, firstly, what the general schema is to which we can reduce each single case of conflict between the opposing solutions to philosophical problems given by different schools of thought, and secondly, how to use such a general schema to resolve the conflict. If so, you are now in a position to apply it to new examples you might encounter and to construct your own solution to such conflicts. The main point was that in each single case we were able to point out a common underlying fallacy shared by both parties, and it was precisely by showing the assumption which both opponents presupposed as harmless, and by revealing the dogmatic character of the assumption, that we were able to settle the dispute and to solve the problem. This presupposition, which we had to extricate from the darkness of the unconscious, can be reduced to the general form of a disjunction or an alternative whose crucial aspect is that its completeness seems to be a matter of logic in accordance with the principle of the excluded middle.

The presupposition can also be reduced to another general form—the assertion that two concepts, or at least their extensions, are the same. In this form both opponents turn out to take the attributes of the objects falling under one concept and transfer them to the objects that fall under the other concept. In this way each opponent produces arguments whose conclusions are incompatible with those of the other. The presupposition amounts to saying that the starting-points of both opponents are irreconcilable. Indeed they seem to contradict each other. The illusion of a contradiction arises from assuming the equivalence of the two concepts. For if two concepts A and B are identical, it is a contradiction to affirm something about A and to deny it about B. But the contradiction disappears if A and B are not identical. The illusion of this identity brings about the conflict between the two parties. Revealing this illusion resolves the contradiction. The illusion arises from swapping two concepts. This concept-swapping operation is facilitated by the misuse of a nominal definition that is arbitrarily introduced. The surreptitious replacement consists in the transition from an analytic judgment (in itself true) to an unfounded or even false synthetic judgment. An analytic judgment asserts the contradiction between affirming A and denying A, whereas a synthetic judgment asserts the contradiction between affirming A and denying B.

If we compare the different examples given, we recognise that this is indeed a *typical* fallacy. For what is usually wrong in philosophical argumentations is that, instead of giving a direct proof of one's own position, they put forward a refutation of the doctrine of one's opponent. We can see how the fallacy works when we consider that the only way to go from refuting the opposing view to the positive establishment of one's own doctrine is by arguing from the same incomplete disjunction that one's opponents use to argue for *their* doctrine—an assumption which is and remains hidden from both parties because they are guilty of the same conceptual or at least extensional identification.

Such an antinomic schema really contains almost all disputed questions which have split philosophical schools into opposing camps and have been debated back and forth so unsuccessfully just because the opponents unconsciously made an unsuspected assumption and only paid attention to producing arguments which were indeed unassailable on both sides. If one has become aware that a presupposition lies concealed behind the nominal definition given, then and only then can one think of questioning it. But bringing this presupposition to light is difficult, precisely because it stems from not paying attention to the difference between two concepts and thus from a conceptual identity that is in no need of argument—the identity of a concept with itself. The only way to draw attention to the difference between concepts is to leave for a while aside all concern for argumentative rigour and to concentrate instead on conceptual clarification, on finding out the content of those concepts we intend to use in our arguments. But you can only see that such work at conceptual clarification is required before worrying about constructing valid arguments if you are not taken in by the seeming logical necessity of a plausible chain of inference, and if you do not close your mind to the insight that the exact opposite chain of inference is equally necessary. But once you have grasped the methodical principle of this whole sophistic dialectic, you will no longer be misled by it.

I now want to leave these considerations behind and just draw your attention to what may well be the most famous relevant example—the 'antinomy of pure reason'. It is more famous than really well understood, for it can also be reduced to our schema, as a particular instance of the logical principle we have been discussing. Because of our time constraints I cannot go very deeply into the antinomy, but I want to say something here for those who know the doctrine. Kant's resolution

Lecture XX

of the antinomy is extraordinarily deep and a work of genius, and one can see that it is all due to his new method of analysis, his critical method; but not everyone can see immediately that it can be reduced to my schema. So I want to explain it a little bit. The two concepts which are here swapped with each other, thus causing the emergence of those antinomies (which according to Kant originate in a transcendental illusion inherent in our very faculty of reason and thus incurable), are NATURE and UNIVERSE.[1]

The cosmological antinomies have to do with the series of conditions that lie before us in nature, i.e. in the field of objects of which we can have experience. Note that NATURE is nothing but

THE SET OF ALL OBJECTS OF POSSIBLE EXPERIENCE,

or as Kant says,

THE SET OF ALL EXISTING THINGS INSOFAR AS THEY ARE SUBJECT TO NECESSARY LAWS

or even

THE SET OF ALL SPATIO-TEMPORAL PHENOMENA

On the other hand, UNIVERSE is the concept or rather the idea (as Kant appropriately says, and remember that ideas are a peculiar kind of concept) of

THE WHOLE OF EXISTING THINGS

or of

EVERYTHING THAT REALLY EXISTS.

UNIVERSE is an idea because it cannot be an object of possible experience, as the resolution of the antinomy explains. The antinomy results whenever we take these two concepts (or at least their extensions) to be the same. Of the UNIVERSE we can say that it consists of all series of conditions. If each single conditioned thing is given, then so are all its conditions, and we have to presuppose the series of conditions to be completed. Yet what we have to presuppose for the UNIVERSE—the completion of the series—in order to avoid an infinite regress cannot be applied to NATURE. Of NATURE we must assume the opposite: all series of conditions are incomplete and incompletable, both in time and in space, both in composition and in division. Hence, an antinomy emerges. On the one hand, we presuppose a whole (the UNIVERSE) and argue that the series of conditions in NATURE is finite and completed; on the other hand, we take NATURE as a NEVERENDING SERIES OF CONDITIONS and argue that the UNIVERSE is incompletable. And so we end up with a thesis and an antithesis opposing each other. The resolution of the conflict consists in showing that the two premises (the UNIVERSE is a WHOLE, NATURE is a NEVERENDING SERIES) are compatible; and this we do by showing that we were assuming that NATURE is the

---

[1] This way of analysing Kant's antinomies is due to Fries (1837, §19). Kant's own exposition and solution is in his *Critique of Pure Reason*, A405–567 B432–595.

same as UNIVERSE. This is the doctrine of transcendental idealism, according to which our knowledge of nature is not knowledge of the universe, not even partial knowledge, for it is just knowledge of the universe *as it appears to us*, as a phenomenon, conditioned by our [epistemic] limitations.

Now the whole of this course of lectures so far has only dealt with the first part of the material we wanted to study, for we have so far only shown that logicist dogmatism (i.e. the attempt to extend our knowledge by means of mere concepts) depends on, and fails because of, one particular fallacy, which we have identified. The illusion common to all such attempts is now clear. It is a case of *quaternio terminorum*, which people incur when they go from an analytic judgment to a synthetic one, and mask the difference by misusing nominal definitions. But, as I said, this was only the first part of our task, in fact the prelude to our proper topic, which was the intuitive philosophy which is in fashion these days. Well then, I would like to argue that we do not need any more tools for this main task, since that intuitive philosophy can surprisingly be reduced to the very same fallacy we have shown to cause the failure of logicism.

You may think this is a paradox, but I am going to prove it, and the proof I am about to offer will then complete the task of this series of lectures. This proof will be the counterpart of the logical or rather logicist dogmatism. Let us call it the open dogmatism in philosophy, open in that there is not even an attempt to give reasons for the principles these philosophers proclaim, for they just place those principles at the beginning and expect us simply to accept them, the presumptuous idea being that anybody who grasps their meaning will immediately appreciate that they are self-evidently true. The principles are therefore taken as axioms, i.e. as immediately self-evident truths, in no need of further argument. That is the specific trait of intuitive philosophy. That is what distinguishes it *prima facie* from logicist dogmatism. Now, I would argue that this is also the only difference between these two ways of doing philosophy, for the open dogmatists are skilled at concealing how they come to their supposedly self-evident principles. When we uncover the crucial moves they hide from view, then it can be seen again that logical dogmatism is the source of their fraudulently self-evident principles. Both schools use in fact the same trick; but the trick is more deeply concealed, there are more veils as it were, and so we have to work a little harder to show that this apparently so different way of doing philosophy is simply a repetition of logicist dogmatism.

What is there in the alleged intuition of the metaphysician? It consists in *Wesensschau*, as it is quite aptly called—an intuition of essences. The intuitive metaphysician does actually claim to 'see' the general essence of those things that philosophy is about—the essence of movement, the essence of nature, the essence of law and justice, and so on. ESSENCE is of course nothing more than an hypostatised or reified GENERAL OBJECT. What one 'sees' is an object; and the objects one 'sees' differ from the objects of ordinary sensory perception only because they are general. An essence teaches us, when we 'see' it, the universal and necessary attributes that belong to individual objects, and they are universal and necessary because they partake in the said general essence, i.e. every specific movement partakes in the essence of movement, every specific law in the essence of the law, and so on.[2]

How does one come to 'see' any such essence? By swapping two concepts and putting their respective objects into one basket. The illusion that it is 'just self-evident' that one is dealing with one and the same set of objects can be explained by the previous assumption that the corresponding concepts are not two but one concept. For if two concepts C and D are identical [if $C = D$], then we can assert without further ado that what is true of the objects that fall under concept D is also true of the objects that fall under concept C, that the attributes contained in D can be transferred to the objects that fall under concept C.

The next step is to elevate such an assertion [namely, $C = D$] to the rank of an axiom that forces itself upon us. It is self-evident that any concept is identical to itself, and it is this self-evidence which (via a misleading argument) gets transferred to a synthetic judgment elevated to the rank of an axiom, a synthetic judgment, in other words, which takes the attributes contained in one concept $[D]$ and transfers them to the objects that fall under another $[C]$. If you do not see where the force of the alleged axiom comes from, if you do not see that it is all due to a fallacy (the swapping of two concepts), and if you believe that you have in front of you a real, self-evident axiom full of wonderful consequences, which anybody will be forced to appreciate the moment it is put forward, then you will, like all other thinkers misled by the same fallacious reasoning, believe that the obvious conceptual identity protects you against each and every mistake in argument. The fallacy works particularly well because the concepts we deal with here lie inside us in an originally obscure and therefore confused shape, and mixed with other concepts. The intuitive philosopher depends on this very fuzziness and on nothing else, and assuming that his fellow human beings have equally blurred conceptions as himself he claims self-evidence for his principles. Yet it is the fuzziness which alone can explain the whole illusion.

Our critical work is, on the one hand, actually facilitated by this way of doing philosophy. For it is so open [about the groundlessness of its claims] that we can immediately see how dogmatic it is. On the other hand, there is also a disadvantage: the replacement of one concept for another is more difficult to uncover and bring to light, precisely because its author wraps it in a cloud, as he does not offer an account of what he had to do to obtain his insight. But with a little practice we succeed in uncovering the trick, and it was in order to give you this practice that I dwelt so long on logicistic dogmatism. With a logicistic dogmatic you do not have to uncover the fallacy by pulling the trick out of the darkness of the unconscious and into the light. You just follow the argument in detail and analyse it in its various steps.

---

[2]Notice the difference between the phrase 'universal and necessary', as used here, and the phrase 'necessary and sufficient', as explained in Footnote 15 in Chapter "Lecture IV". Nothing could be farther from an extensional definition of the sort Nelson is only interested in than a 'statement of the essence', as is common to Aristotelian and intuitive metaphysicians (two schools of thought otherwise quite different).

Let us use a concrete example to illustrate this general point—Brentano's definition of the GOOD, which I mentioned earlier on (Brentano 1889; see Chapter "Lecture XIV"). By means of this definition the proponents of the philosophy of the *Wesensschau*, of the intuition of essences, believe they have put ethics on a new scientific basis. You remember the definition. It read:

(1) We call something GOOD if LOVING IT IS RIGHT.

I have already shown how a futile and empty statement (1) is, and that it says nothing more than that something is GOOD if it is THE OBJECT OF A LOVE THAT IS RIGHT, which is just a convoluted roundabout way of saying:

(2) We call something GOOD if it *is* GOOD.

For a love is right if it is directed at something that deserves to be loved and thus actually good. This wisdom has us moving in circles. For how should we know, according to this definition, whether something is good or bad? By loving it with a right love, and the rightness of this love consists in being directed to something that is good. When is love right? The puzzle is solved by observing that right love is self-evident. The self-evidence is immediate. What we have here is direct self-evident knowledge of the essence of the good.

The circularity of Brentano's definition is not its only defect. The fallacy goes deeper. The definition is supposed to explain the fundamental concept of ethics. The moral good is what we are talking about here. However, the concept of the MORAL GOOD is in fact different from the concepts LOVELY, ESTIMABLE, and in short VALUABLE. If we love a flower because it is beautiful, if we confer an aesthetic value on it, we cannot possibly be talking of the good in the ethical sense. The moral good is surely something about our own behaviour. If we apply Brentano's definition to our own behaviour, we might then say that if our own behaviour is the object of right love, then it is morally good. The listener who has not been obfuscated by a sophistic philosophy will not easily accept what follows from such a statement. He is sufficiently aware from his own experience that our behaviour, at least in the first instance, is morally good only if we fulfil our duty, only if our behaviour agrees with our duty; and he also knows from experience that duty is anything but lovely or pleasant. Duty does not give preference to an action above an alternative action because it is lovely. What should such loveliness rest on? If we for example control our inclination to take other people's possessions, it would be difficult to see what was so exceedingly lovely about this action.[3] Or if we fulfil a promise, where fulfilling it has no other value than to fulfill our duty, it is difficult to see what is supposed to be so lovely about this action. We have only fulfilled our duty and

---

[3] This argument goes to the heart of many claims made in moral philosophy, all the way from Plato's Socrates (whose appeal to 'the beautiful' or *tò kalón* so confuses his interlocutors) down to the Romantics and, of course, the thinkers Nelson is discussing here. To make Nelson's argument even sharper, the reader may think of those occasions in which fulfilling one's duty leads to an ugly action. This state of affairs is not at all uncommon within a certain subclass of moral dilemmas.

obligation, there is no merit here, no particular value that we might claim for our behaviour; we have only fulfilled the minimum, in the absence of which our behaviour would be worthy of contempt. The dogmatist ignores all this, and he ignores that his allegedly self-evident statement so obviously contradicts common sense, because in actual fact he does not hold on to what is self-evident to him. Instead, he starts from a purely conceptual study of the problem, and he gets entangled in ruminations and labyrinths which he cannot get out of because he lacks a guideline for the proper solution of the problem, so that he falls victim to any fallacy that comes his way. And yet the fallacy forces itself so strongly on him that what comes out of it really appears to him to be a self-evident statement.

I would like to read to you just a few sentences written by one such intuitive moral philosopher. I shall take them from a treatise by Max Scheler, who is held in high regard nowadays: *Formalism in Ethics and Non-Formal Ethics of Value* (1913). At the beginning of this treatise the following axioms are set forth[4]:

- Good is the value in the domain of the will which is inherent in realising a positive value.
- Bad is the value in the domain of the will which is inherent in realising a negative value.

The system of a supposedly scientific ethical doctrine is then dogmatically founded on these axioms. I shall have a closer look at this ethical doctrine next time.

References

Brentano, Franz. 1889. *Vom Ursprung sittlicher Erkenntnis*. Leipzig: Duncker & Humblot. [English translation:. *The Origin of the Knowledge of Right and Wrong*, Westminster, Archibald Constable & Co., 1902].

Fries, Jakob Friedrich. 1837. *System der Logik: ein Handbuch für Lehrer und zum Selbstgebrauch* [A System of Logic: a Handbook for Teachers and Independent Study], 3rd ed. Heidelberg: Winter. [First edition 1811].

Scheler, Max. 1913. Der Formalismus in der Ethik und die materiale Wertethik (mit besonderer Berücksichtigung der Ethik Immanuel Kants), I. Teil. *Jahrbuch für Philosophie und phänomenologische Forschung* 1: 405–565. [English Trans. *Formalism in Ethics and Non-Formal Ethics of Value: A New Attempt Toward the Foundation of an Ethical Personalism*, Evanston (IL), Northwestern University Press, 1973].

---

[4]See Scheler (1913, Chap. I, §2, Axioms II-1 and II-2). The German title of Scheler's book would be literally translated as *Formalism in Ethics and Material Ethics of Value*, where the word 'material' refers to 'content' as opposed to 'form'. I follow the decision of the translator of this work into English, since it is true that 'material' would have connotations alien to Scheler's intentions.

Lecture XXI

Abstract Intuitive (or intuitionist) philosophers seem so different from, indeed opposed to, logicists that it is easy to think the mistakes in their arguments would be equally different. In fact, intuitive philosophers (like Max Scheler and other 'phenomenologists') seem not to argue at all but just to communicate their visions to the world. A careful analysis, however, shows that they do argue and in their (mostly implicit) arguments fall into the exact same concept-swapping fallacy by which synthetic judgments (about the 'essences' and 'values' they are able to apprehend directly) are unconsciously derived from analytic ones.

By comparing several examples chosen at will we have arrived at a general theory of the fallacies committed by logicism. One may think that such a theory, although in itself interesting and instructive, does not have many applications or results to offer, since it only has to do with the technical logical connection of the propositions that make up a [philosophical] system. It appears that such a system interest us first of all because it contains true propositions, not so much because the formal connections between those propositions are technically correct so that the whole system is logical. For even if there are formal errors, if a proposition is derived from another through a fallacious inference, only logicians would be interested in pointing to the fallacies. There does not seem to be any other but this purely logical interest, for we know that from fallacious inferences and false premises we can still (and if we are lucky we do) draw true results. It seems that what we may call the 'material' import of a philosophical system lies entirely in its individual propositions and particularly in its principles; and since these principles qua principles cannot be logically reduced to any other propositions, so we cannot criticise them via the theory of fallacies developed in this course.

This is wrong. Our theory of fallacies proves its mettle by showing that its scope is wider than it seemed when we were working it out. I have shown this in my last lecture already. If the alleged axioms placed at the start by a metaphysician in order to erect his system upon them spring, like Minerva, from Jupiter's head, so that we cannot criticise the method used to arrive at them, if we are forced to be modest and admit we lack the higher intuition that would give us access to that metaphysical truth he proclaims, then we would have to give up and leave the business of building a philosophical system to him who is endowed with such intuitive powers. But this is not so, for as we went deeper into it, we found the alleged axioms to be smuggled in by a form of reasoning which after doing its work vanishes into the darkness of the unconscious, so that the author himself does not know any more how he arrived at these propositions which he claims to have obtained by intuition. Only by digging a bit deeper can we retrace this path and discover that it consists of a swapping of concepts of exactly the same kind we found that all fallacies of logicism go back to.

The example I began to discuss is one of the most important and far-reaching in the entire history of philosophy. We can call it 'ethics of goods'. By that I mean any ethical doctrine which reduces all moral demand to a doctrine of value, so to speak a doctrine of goods. Something is a good insofar as a value is ascribed to it. So an ethics of goods is about reducing the doctrine of duties to the doctrine of the value of things. This ethics enters stage as a system which seems to be based on self-evident principles, on axioms. Last time I read you a relevant passage from a highly praised treatise by Scheler (1913), entitled *Formalism in Ethics and Non-Formal Ethics of Value*. The passage gives us the axioms[1]:

- Good is the value in the domain of the will which is inherent in realising a positive value.
- Bad is the value in the domain of the will which is inherent in realising a negative value.

I have already tried to show how an obvious course of reasoning does indeed lead to this issue. For if you think about the issue in a purely conceptual manner and using Scheler's axioms, then you will be deceived into thinking that the reason why we prefer to do our duty over any other alternative action must be that the dutiful action exhibits a higher value. To obey duty would be an utterly blind act if we did not see a reason to choose to do it in preference to all other acts. So the preference given to duty is based on the higher value of the dutiful action.

I pointed out before that this idea runs against common sense, and by common sense I mean the sound judgment of people who are not seduced by a sophistic spin. For such people have had experience of many cases in the ordinary course of life to know that people do not decide to fulfill their duty because they just love it. The action that is identified as one's duty is put first not because of the value we discover it has and that makes it lovely [recall the definition of GOOD in Brentano (1889)] but because it *is* one's duty, because it is what is *required* of us in a way that commands our *respect*, even if it should be ugly, even if we are reluctant to do it in view of what we know about the worth and unworth of things. The curious thing about duty is precisely that it commands us to overcome our reluctance. Now if somebody sees this, if somebody sees that the allegedly higher value is an artefact

---

[1]See Scheler (1913, Chap. I, §2, Axioms II-1 and II-2). In the German original Scheler literally speaks of 'material' values as opposed to 'formal' ones. The English translator, aware that the expression 'material value' might be misunderstood in the sense of vulgar or sophisticated 'materialism', has chosen to replace 'material' by 'nonformal'. I follow this convention here.

to which duty is supposed to be reduced, then he will be tempted to continue this line of argument to its logical end and so to reject the alleged duty. He is forced towards the view that the idea of duty has no support whatsoever, that it is a blind idea, irrelevant for a moral philosopher who has a higher insight, so that he must banish it from his ethics.[2] He leaves the idea of duty for lesser men, who are unable to rise to the higher ethical standpoint. He has so to speak overcome duty through love.

The same assumption (that duty is to be chosen because of the higher value of the required action) leads thus to two different argumentations and two different kinds of distorted ethics, one the mirror image of the other. One side argues that, since we are commanded to do our duty, and therefore doing it is a necessary condition of the value of our action, the fulfilment of duty is already enough to give merit to the action. The other side argues that, since fulfilling our duty is not enough to enhance the value of an action, it is not necessary to do our duty. What do we have here? The assumption that if this condition of value is really necessary, then it must also be sufficient. A necessary condition is equated with a sufficient one, i.e. a necessary condition of value is equated with a sufficient condition of value. On one side of the dispute we have a premise according to which the fulfilment of duty is a condition of the value of the action, and from this premise the conclusion is drawn that the fulfilment of action already grants it a positive value, one that surpasses all others. On the other side of the dispute we have the premise according to which the fulfilment is not sufficient to give value to the action, and from this the conclusion is drawn that the fulfilment of duty is not even a necessary condition of the value of action. We can put both premises together—'Fulfilling one's duty is a necessary condition for the action to have a value', and 'Fulfilling one's duty has in itself no value'. There is no contradiction between them, and so we only need to drop the false assumption that if fulfilling one's duty is a necessary condition of value, then it must also be a sufficient one (see Fig. 1).

Scheler's treatise does not clearly take sides in this dispute; it fluctuates between them. Sometimes one argument is foregrounded and sometimes the other. But it will be interesting to have a look at a few more passages.

Max Scheler says[3]:

> The first essential fact is that all values (whether they are ethical, aesthetical, or whatnot) are divided [...] into *positive* and *negative*. This belongs to the *essence* of value [...].

This is Scheler's thesis, and it sounds extremely plausible if, as I said, the issue is just a matter of ruminating and playing with concepts. To repeat, the author states that in every valuation the possible values can be divided into positive and negative,

---

[2] This is exactly what happened in modern aestheticism, as exemplified by the Bloomsbury group, to which the likes of Bertrand Russell and John Maynard Keynes (inspired by Moore's *Principia Ethica* of 1903) belonged. For an exquisite dissection of this ethical tendency and the regrets that come with maturity of thought see Keynes (1949).

[3] These and the following quotations within this and the next paragraph are taken from Scheler (1913, Chap. II, Sect. B, §1).

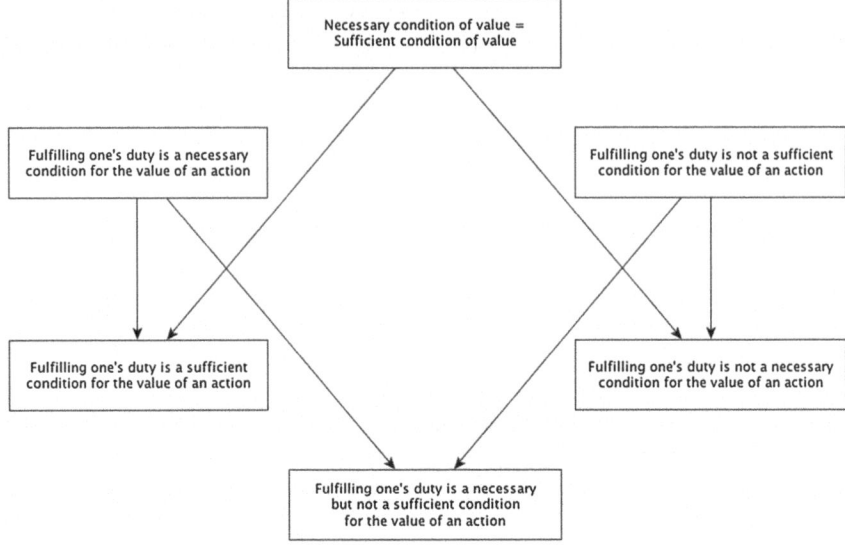

Fig. 1 Fulfilling one's duty is necessary but not sufficient for an action to have value

and it is explicitly added that this is also true of ethical values. In saying this the author loses sight of the idea that fulfilling one's duty has no positive value in itself, that there is no positive moral value but only a negative one—the moral unworthiness of the breach of duty, for fulfilling one's duty is mandatory not meritorious. He forgets this because he does not focus on his ordinary judgment, surely to be present in any particular case; and so he comes to state

> that every 'ought' must be based on values, *i.e.* that 'ought' only applies to values..., as well as... that positive values ought to be and negative values ought not to be.

So the moral law is reduced to values by Scheler, contrary to the idea that moral valuation is everywhere negative.

In another passage Scheler says that

> every nonnegative value is positive,

and he adds that

> the value *good* [viz the exclusive predicate of moral action] is just a *domain of application* of the formal laws of value,

so that

> any application presupposes the values *good* and *bad*,

and

> the said laws rest on *intuitive* essential connections.

The author is thus saying that he owes all these discoveries to an immediate knowledge of the essence of the good [the 'essential connection' he is able to apprehend directly]. The true [logical] relationship—brought to light by our conceptual analysis—is thus inverted. Scheler uses these 'self-evident' axioms to reject Kantian ethics, to dispute the Kantian thesis that the moral law must precede every decision as to what is good and bad, and not the other way around, that it is based on first deciding what is good and bad.[4] Scheler maintains that Kant's attempt to reduce the meaning of the words 'good' and 'bad' to the content of a norm fails because Kant does not show that without a norm there is no such thing as good or bad. He rejects Kant's idea that the meaning of the words 'good' and bad' is utterly exhausted by whether an action is according to the law or against the law.[5] Never mind the sheer impertinence of all this. I just ask you to consider carefully what it is that you feel when you feel that something is bad or that something is good. For this is what Scheler has not done, he has never looked at his own feelings of good and bad, preferring instead to abandon himself to an alleged intuition of all those things that spring out of his conceptual spinning, and which (thanks to the fallacy we have explained) impose themselves upon him with such force that he feels compelled to claim an immediate intuition of them.

I have quoted from Scheler's treatise in the *Yearbook for Philosophy and Phenomenological Research*, edited by Edmund Husserl, volume I. Next I will quote from 'The Idea of a Moral Action' (1916) by Hildebrand, a monograph published in vol. III of the same journal[6]:

> Any consciousness of duty that is not *secondary*, that doesn't flow from a nonformal value, must appear to us as *blind*. It is true that 'Thou oughtest' has formally a categorical character which utterly distinguishes it from any arbitrary coercion or command. Nonetheless, it remains unwarranted unless it flows from an awareness of a nonformal value. Only when it is added as an 'epiphenomenon' to a nonformal valuation, can any consciousness of duty be considered to be clear and the basis of a nonblind voluntary response, and even then the will is responding to the value itself, so that the consciousness of duty is just an aid that can help in particular cases.

Hildebrand speaks [elsewhere] of duty as

> an imperative floating in the air [that] in itself is either always blind or else leads directly to a rejection of a categorical response to values by changing one's attitude.

Max Scheler for his part argues in another passage[7] that the consciousness of duty

---

[4]On Kant and his thesis that 'good' and 'bad' are determined by the moral law, see Appendix below, especially Footnote 11.

[5]See Scheler (1913, Chap. I, §2).

[6]The first quote comes from page 232 and the second from page 236 of von Hildebrand (1916).

[7]See Scheler (1913, Chap. II, Sect. B, §5). The original text is somewhat ambiguous between 'the consciousness of duty only enters stage when people lack a full moral insight' and 'the consciousness of duty only enters stage when people lack moral insight in the proper sense', which is of course Scheler's sense.

only enters stage when people lack a full moral insight.

The obligatoriness of norms is [for him therefore] just a matter of values[8]:

> Therefore 'duty' can only become what is *good* or what, *because* it is good... necessarily 'ought' to be.

It is [according to him] an insight into the

> a priori structure of the realm of values which... entails the 'necessity' [he means obligatoriness] of a norm. In contradistinction, putting that necessity (not to say 'duty') *before* the insight into what is good,... is false.

I don't have enough time to go deeper into this otherwise very important doctrine that ethics would be about goods; but I would like to clarify what my whole point is. There is an illegitimate transition from a perfectly self-evident analytic proposition to the synthetic assertion contained in Scheler's axioms, a transition he himself is utterly unaware of. This much is hopefully clear: the moral value is predicated of an action *because* it is a duty, and not the other way around. For the character of duty that an action has is *not* derived from the supposed value of the action. We can only call a moral action 'good' *because* it is required and so to be preferred over all other alternatives. It is not true that the action be required because it is good or estimable. To assume the latter is the fallacy of any ethics of goods, i.e. of all attempts to derive the doctrine of duty from the doctrine of the good.

Now what is the source of the illusion of self-evidence that leads to all such attempts? 'It is our duty to do good'—this statement sounds undisputable. It seems that all one has to do is to call to mind the essence of the good and one will immediately see that the statement is true. It seems that all one has to do is to understand the meaning of the sentence. Yes, nothing can be more self-evident than that it is our duty to do good. The dialectical illusion rests on the ambiguity of the word 'good'. On the one hand, the word 'good' means [the same as] 'morally good' or, to get rid of all ambiguity, just 'moral'; on the other hand, it means [the same as] 'valuable'. Yet the meaning of the statement above changes a lot depending on whether we understand 'good' to mean one thing or the other.

If by 'good' we mean 'moral', i.e. the fulfilment of duty, then the statement above is trivial, it only expresses an analytic judgment—that duty can only be fulfilled through the fulfilment of duty. For to say 'this is my duty' means nothing other than 'I have to fulfill this'. Things are completely different if we take the meaning of 'good' to be 'valuable'. In this case the statement above expresses a synthetic judgment, namely that we ought to do what is valuable or what has a higher value. This statement can never be reduced to the other statement, which is analytic and indeed self-evident. The analytic statement does not contain the concept of VALUABLE. The synthetic proposition, which is what the fuss is all about, could only be received as true if, in addition to the concept of GOOD in the sense of VALUABLE,

---

[8]This and next quote from Scheler (1913, Chap. II, Sect. B, §7).

we would claim to have an intuition of the essence of the good according to which it is a duty to do what is more valuable.

Two concepts are being confused here then, GOOD in the sense of MORAL, and GOOD in the sense of VALUABLE, and only by replacing one with the other can one claim that the axiom is self-evident. The illusion of self-evidence is transferred from the really self-evident analytic statement to what is in truth a synthetic proposition, unfounded and indeed false. The analytic statement is confused with the synthetic false one because of the ambiguity of the word 'good', so much so that the latter not only seems to be true but self-evident to boot. Of course, this 'self-evidence' vanishes before common sense, whose judgment of every single case in which we have to fulfill our duty is pretty firm, as pithily expressed by Wilhelm Busch [in his poem 'Pious Helen']:

> The good, and this is true,
> Is the bad one did not do.

The whole pompous endeavour of the ethics of goods is wrecked by this homely truth.

No more examples will then be needed to illustrate that doing philosophy in this allegedly intuitive manner, if indeed the direct opposite to logicism, rests on the same fallacy, by which it smuggles an appearance of self-evidence and the possession of metaphysical intuition. The doctrine of metaphysical intuition and self-evidence of axioms is just so much embellishment and prop that masks the underlying logicist dogmatism and gives the edifice a new glimmering—even seducing—façade. Anyone who has identified the fallacy once will not be deceived by the façade and he will always recognise the fundamental disease afflicting logicist dogmatism.

For the remaining time of this course I intend to move on to a different point, to another typical fallacy which accompanies and reinforces the one we've been discussing. This second fallacy occurs both in logicist dogmatism and in the open dogmatism of 'intuitive' metaphysics, both in mystical and in scholastic philosophy. Mysticism is nothing more than the pretension of being able to *intuit* what other mortals can only *think*. We will better understand the recurrence of the mystical illusion in the history of philosophy if we take this second fallacy into account.

We were able to explain the origin of the first fallacy in terms of the neglect of the distinction between analytic and synthetic judgments, which is an important defect in the logical doctrine of judgment; and almost the same is true of the second fallacy. For this second fallacy stems from ignoring the doctrine of the logical form of judgment. Such ignorance proves to be the rock-bottom explanation for the monstrous metaphysical derailments we see in the history of philosophy. The elementary logical error I have in mind is revealed in the distortions and in the fallacies committed on a grand scale by a particular philosopher when obtaining his metaphysical results.

In Kant's *Critique of Pure Reason*, under the modest title of an 'appendix' to one of its main sections, one can find a remarkable and subtle doctrine of which few people have noticed more than the rather opaque title. I mean the appendix to the 'Amphiboly of the Concepts of Reflection'.[9] By a further application of his critical method was Kant the first to uncover here the fallacy whose corrupting role I want to tackle. Although he only approaches it from a particular angle, anyone who has but once grasped what Kant says there about this fallacy will easily grasp its general significance and be in a position to apply it elsewhere. The limited target of Kant's subtle critique is Leibniz's metaphysics, and in fact he only introduces this discovery as a means of identifying Leibniz' fallacies at the deepest level. He succeeds indeed in reducing the whole drift of Leibnizian metaphysics to one simple logical error. Kant's discovery of that error is at the same time the discovery of a fundamental mystery—that it is the *form* of the error that accounts for its seductive power—even if in his text Kant only explores the particular form the error exhibits in Leibniz's system. I will talk of this error in my next lecture.

References

Brentano, Franz. 1889. *Vom Ursprung sittlicher Erkenntnis*. Leipzig: Duncker & Humblot. English translation: *The origin of the knowledge of right and wrong*, Westminster, Archibald Constable & Co., 1902.
von Hildebrand, Dietrich. 1916. Die Idee der sittlichen Handlung [The idea of a moral action]. *Jahrbuch für Philosophie und phänomenologische Forschung* 3: 126–251.
Keynes, John Maynard. 1949. My early beliefs. In *Two memoirs*. London: Rupert Hart-Davis.
Moore, George E. 1903. *Principia Ethica*. Cambridge: Cambridge University Press.
Scheler, Max. 1913. Der Formalismus in der Ethik und die materiale Wertethik (mit besonderer Berücksichtigung der Ethik Immanuel Kants), I. Teil. *Jahrbuch für Philosophie und phänomenologische Forschung* 1: 405–565. English translation: *Formalism in Ethics and Non-Formal Ethics of Value: A New Attempt Toward the Foundation of an Ethical Personalism*. Evanston: Northwestern University Press, 1973.

---

[9]See *Critique of Pure Reason*, A260–292 B316–349.

Lecture XXII

Abstract There is a subclass of intuitive philosophers that indulges in a kind of mysticism. Mystical philosophers (e.g. Spinoza, Hegel, Spengler) commit the same concept-swapping fallacy that has been discussed in this book, but in their case it is additionally sustained and fed by a glaring elementary logical error: replacing the ordinary predicative logic which Aristotle founded by a pseudo-logic of identity in which the distinction between concepts and things—what we say and that of which we say it—is rejected in favour of empty formulae in which concepts are said to be identical whilst at the same time different. Finally, it is shown that the kind of analysis pursued here leads to a new clarity on a general philosophical predicament: all forms of metaphysical dogmatism share the same prejudice as their arch-enemy, metaphysical scepticism.

If one focuses on the most general [cognitive] trends dividing philosophical schools, then it is possible to reduce the whole division to a conflict about the very first principles of logic. It can be argued that people do philosophy according to two different types of logic. One logic we may call Aristotelian or rather *Aristotelian-Kantian*, the other we may call Neoplatonic or mystical.[1] The first and foremost difference lies in abstraction, generalisation, concept-formation. The most simple and obvious way to abstract and build concepts consists in classification, putting in a class certain objects of perception under one general idea. The things thus classified have, as elements of the class, a logical relationship on the strength of the concept that classifies them.[2] The logical relation between a class and its elements is

---

[1] The opposition between 'Aristotelian-Kantian' and 'Neoplatonic-Fichtean' logic, which Nelson will sketch here, is more fully developed in another course of lectures, that has been translated into English (see Nelson 1962). It is clear that Nelson is convinced that Kant's modifications to traditional Aristotelian logic were very important, a question that is moot, to say the least, from the point of view of contemporary logic. Two interesting positions on this question that it is worth the reader's while to compare are Hintikka (1973, 1996) and Bar-Am (2008).

[2] Compare Nelson's sketchy description with Russell's expansive one (1903, §§20–26, 48–55, 66–79).

quite different from that between a whole and its parts.³ The latter is a *real* relationship not just a logical one. The *elements* of the class [genus, species] are subsumed under it, whereas the *parts* are actually united in the whole and united to each other.

These are two different abstracting procedures—in one we go from the individual to the class [species or genus], in the other we go from a part to the whole it belongs to. Jakob Friedrich Fries was the first to make this distinction in his *Logic* (1837, §19). He called the former 'qualitative' and the latter 'quantitative' abstraction. Well then, what characterises mystical logic is that it does not recognise this distinction. It ignores the logical relation of subsumption between a member of a class and the class it belongs to, and only admits the integration of individual things into an *essence*, which is just a hypostatised [or reified] concept. This 'essence' is very similar to a whole in the sense of Fries' quantitative abstraction. Such wholes, however, are obtained by generalisation or qualitative abstraction. Some people talk here of 'the *type* of phenomena', and they see this *type* as their real being.⁴ Individual things are unified in the *type* or *essence*.

This entails an inversion of the subject-predicate relation in the categorical judgment. According to Aristotelian-Kantian logic, a categorical judgment, like (1):

(1) This board is black

is such that an object [denoted by 'this board'] is subsumed under [or classified by means of] a concept [BLACK], the copula 'is' being the means to express this subsumption. According to mystical logic it is precisely the other way around—the concept is hypostatised as an essence, and the individual is just one of the parts or manifestations of the essence, so that the hypostatised concept [BLACKNESS] takes on the role which in Aristotelian-Kantian logic corresponds to the real objects of intuition. By contrast, the objects themselves are analogous to Aristotelian-Kantian predicates and become mere attributes of the concept ['this board' becomes the attribute TO BE A BOARD]. In this manner, mystical logicians believe that they come to know what the individual objects of intuition really are, how they are really constituted, by integrating them into the hypostatised concept that encompasses them all [by knowing the essence of a board they already know what makes boards what they are without having to study them in the real world]. So mystical logicians need no induction and no research of natural laws in order to gain knowledge of the real constitution of things given in sensory intuition, as demanded by Aristotelian-Kantian logic. Mystical logicians make their discoveries in a

---

³The whole-part relation has caused innumerable headaches. The most famous logical development of the whole-part relation goes from the very crude beginnings in Husserl (1901) all the way up to the sophisticated 'mereology' of Leśniewski (see Luschei 1962; Urbaniak 2014). The technical level of discussion in this lecture is very far from such developments yet hopefully sufficient to identify the fallacies in question.

⁴This idealistic use of the word 'type' (in a sense similar to Plato's *eidos* or *idea*) is characteristic of certain German thinkers, though not all—thus Max Weber's use of *ideal types* is quite different; see e.g. Ringer (1997).

Lecture XXII

nonmediated way—they look at things and put them together under the canopy of the hypostatised concept.

My first example, and a very famous one, will be Spinozism. His problem was body-mind dualism, or to use Spinoza's [Cartesian] terminology, the dualism of *res extensa* and *res cogitans*. The unity of the world can only be known if we succeed in overcoming this dualism. That is what Spinoza believes his metaphysics can do. How? By equating *res extensa* and *res cogitans* as [instances of] Being. For *res extensa* IS and *res cogitans* IS as well. So the concept of BEING is their superordinate concept; it suffices to hypostatise the concept into an essence and an ultimate ground of both *res extensa* and *res cogitans* in order to overcome dualism.

We thus arrive at postulating one single existing absolute substance. *Res extensa* [falsely] looks as though it were absolute, and so does *res cogitans*. Spinoza thus erects a transcendent metaphysics that rises above all experience by using nothing but concepts. It sometimes takes the form of logicist dogmatism, when the author carries out his task consciously; and sometimes it takes a mystical form, the form of an apparently intuitive metaphysics, when what Spinoza is doing vanishes into his unconscious and he just puts the result at the start of his system as though he had 'seen' it intellectually, as though it were an immediately perceived, self-evident truth. We see that the distinction between logicism and mysticism, in itself quite substantial, becomes superficial when viewed from the logical perspective we are taking here. The fallacy of both philosophies being at bottom the same fallacy, the difference between them becomes irrelevant.

There is a different way to analyse the fallacy. Leaving concept-formation aside, let's consider judgment as such. I said earlier that in Aristotelian-Kantian logic the form of a simple categorical judgment consists in subsuming an object under a concept. For that to be possible objects have to be given to us originally in sensory intuition. Reference to the object (the very function of the subject) is realised by ostension, by words such as 'this'. By such words we express that we do not mean the *concept* BOARD but rather that we are *referring* to a particular object given in sensory intuition. That is what the word 'this' does. This ostensive work of the judgment (taught to be necessary by Aristotelian-Kantian logic) is ignored by mystical logic. Their judgments, if one can still speak of judgments here, have the form (2) [instead of (1)]:

(2) BOARD is BLACK.

The word 'is' loses here the meaning it has in Aristotelian-Kantian logic. In mystical logic this word means [the same as] 'equals' [like in (3)]:

(3) BOARD = BLACK

Instead of judgments we have here a comparison, a conceptual formula which equates or distinguishes concepts.[5] But judgments are not about concepts being

---

[5]The German word for 'conceptual formula' is *Vergleichungsformel*, literally 'formula of comparison'. Nelson explains what he means as follows: "[For the mystical logician] an expression

equal to each other [in some respect] or different from each other [in some other respect]; they are about the real constitution of things.⁶ If we would use these words in Aristotelian-Kantian fashion, we would have to write rather (4):

(4) BOARD ≠ BLACK

for the concept BOARD is different from the concept BLACK. But in mystical logic sentence (1) has another meaning, according to which there is no problem in saying that BOARD = BLACK, like in (3). The law of non-contradiction fundamental in Aristotelian-Kantian logic does not apply in mystical logic. One can place an affirmative judgment side by side with the corresponding negative one without problem as equally true. For instance, one can say that

(5) GOLD = SILVER,

meaning that gold and silver are the same, which of course they are insofar as they are both metals; and then again one can also say that

(6) GOLD ≠ SILVER,

for they are also different, say in colour, specific weight, and so on. Gold is both silver and not silver. The law of non-contradiction does not apply in this logic.

My second example had already come up in an earlier lecture. One of the most distinctive modern exponents of non-Aristotelian logic is Johann Gottlieb Fichte. One could almost say that Fichte codified non-Aristotelian logic the way Aristotle codified the classical logic that goes under his name. One of Fichte's non-Aristotelian doctrines is the 'general self' [see Chapter "Lecture XII"].

(Footnote 5 continued)
like '$a = b$' means nothing else than 'the two things, a and b, are in some respect equal', which of course does not exclude that they are in some other respect different." Analogously for expressions like '$a \neq b$'. See Nelson (1962, 513). In other words, a conceptual formula is any form of words in which something is asserted to be vaguely equal or inequal with something else. The word *Vergleichungsformel* stems from Fries (1842); Kant himself only wrote of *Vergleichungsbegriffe*, 'concepts of comparison' (*Critique of Pure Reason*, A170, B216; cf. Fries 1828, §57, 1837, §19).

⁶This assertion is hugely important to understand Nelson's brand of critical philosophy. Judgments, whether scientific or metaphysical, are about real existing things and their real natures and relations, whereas the critique of reason produces statements that are not about reality but about our concepts. Mystical pseudo-logicians would confuse these two levels. We would today say that Nelson (like Fries before him) was an extensionalist in logical matters. Due to the fact that modern mathematical logic only entered German academia after Nelson's death (with the publication of Hilbert and Ackermann 1928), he found it difficult to put across his meaning. Like Frege (1892), Nelson uses the traditional terms *Begriff* ('concept') and *Gegenstand* ('object') to convey the difference which in first-order extensional logic is represented by predicate letters and individual variables. It has been the merit of Hintikka (1973, 1996) to use modern logical notation to give clear expression both to the Kantian distinction between analytic and synthetic, and to the corresponding links between 'concept' and 'object', 'thinking' and 'intuition'. Basically, the trouble with Hegel and other 'mystical logicians', according to Nelson and Fries, is that they improperly and confusedly want to 'intensionalise' all sentences, even when they are clearly extensional. For a different view of what Hegel was after, see Bencivenga (2000).

Lecture XXII   195

How does he arrive at such hypostatisation? Fichte takes the concept of SELF, which in Aristotelian-Kantian logic means just the class of individual minds. But mystical logic deviates from that. In mystical logic this concept is viewed as an essence, as the ultimate ground of every single mind, all of which are unified in it. In this essence all individual minds are identical, each one being a self:

(7) All individuals are the same insofar as they are selves.

This conceptual formula is of course correct [as far as it goes], but Fichte surreptitiously replaces it with a false judgment that all individuals are, as a matter of reality [not just conceptually], and losing their respective individualities, unified in the absolute Self. That's how one comes from a conceptual formula to a metaphysical axiom.

The same example will help illustrate the use of the negative version of that conceptual formula. I have earlier spoken of the following proposition [see Chapter "Lecture XV"]:

(8) The thinking subject is opposed to the object thought.

No identity is stated in (8), but rather a difference. Again, it is a correct conceptual formula. Thinking *is* different from thought. The two concepts (THINKING and THOUGHT) *are* different. But if one mistakes the conceptual formula for a [proper synthetic] judgment, then one arrives at the false metaphysical statement that self-knowledge is impossible. For self-knowledge is defined by saying that the subject and object of knowledge are identical; but if they are conceived as opposites, then the correct yet empty conceptual formula leads (when mistaken for a judgment) to denying that self-knowledge is possible at all.

This fallacy is in fact committed by Fichte, who draws the conclusion that nothing can at the same time think and be thought. What falls under one concept cannot fall under the other concept, and so, from a different side, he again comes to the same conclusion, namely his absolute Self. For this absolute Self is that subject which cannot be object, and which we have to assume in order to understand how the individual self can be object of knowledge. The individual self not being able to know itself, a non-individual subject is required to know it, if one argues [in this Fichtean way] from the following conceptual formula:

(9) The thinking subject is opposed to the object thought.

A third example is the famous Hegelian statement[7]:

(10) The rational is the real, and the real is the rational.

This statement makes a trivial conceptual formula [RATIONAL = REAL] look as if it were a profound metaphysical assertion, with the additional advantage that appealing to it can quash all criticism of the status quo. By this maneuvre (arguing that what is rational is real and what is real is rational) Hegel's philosophy of law

---

[7]See Hegel (1821, Preface).

turns out to be a radical apology of the status quo of his time—the Prussian police state in the end.

Oswald Spengler is the modern proponent of this metaphysical jugglery, whose popularity—a fascinating phenomenon—rests solely on his ability to play this trick like a virtuoso.[8] He has put forward a new metaphysics of history on the basis of the conceptual formula (11):

(11) All cultures are the same.

They *are*, of course, insofar as they all fall under one general concept—CULTURE. Spengler uses this trivial analytic proposition to introduce surreptitiously the metaphysical judgment that all cultures follow the same course, meaning that if you know the course of one culture, then you con prophesy the course of all future ones. The general concept of CULTURE is here hypostatised into a proper object, the essential type of all particular cultures, the *Urphänomen* of cultures. The single cultures become mere manifestations [avatars] of an underlying ur-essence of culture. Aristotle had already unmasked the fallacious maneuvre used here. His example[9]:

Koriskos is a man
Socrates is also a man
___
Koriskos is Socrates

Koriskos and Socrates are the same, i.e. insofar as they both fall under the concept MAN. The fallacy consists in drawing a metaphysical assertion, i.e. that these two men are one in reality, from their having both the attribute of being a man. If he had known the bestsellers of our time, Aristotle could just as well have chosen Spengler's statement that being Prussian and being socialist are one thing[10] and modify the example accordingly:

Being Prussian is a matter of power
Socialism is also a matter of power
___
Being Prussian is being socialist

But do modern logicians *literally* endorse this conception of judgment at the basis of mystical logic? I want to show they do [and that I am not attacking a straw man] by means of an example. Natorp states in his *Logic*[11]:

___

[8]See Spengler (1918). Nelson considered the cultural phenomenon of that author's popularity and influence so important that he dedicated a whole book to its analysis (Nelson 1921).

[9]Not in this form in Aristotle. The nearest would be *Sophistical Refutations* 166^{b}32–36.

[10]See Spengler (1920, 97–98).

[11]See Natorp (1910, 21). By 'modern logicians' Nelson meant those German professors who during the first two decades of the twentieth century kept writing books on 'logic' whilst ignoring the contemporary development of mathematical logic. Please remember that the first edition of the authoritative treatise on the subject by Hilbert and Ackermann, which changed the German situation forever (although not at once), only appeared in 1928, a year after Nelson's death.

Lecture XXII

The negative judgment, '*b* is not *a*', means a judgment of difference (second stage of quality), as the positive [or affirmative, '*b* is *a*'] means the simple assertion of identity (first stage).

It is a pity that my time is running out and I can't elaborate on this fallacy a bit more. Kant called it 'the amphiboly of the concepts of reflection', i.e. of purely logical concepts, by which we, for instance, compare objects or concepts as identical or different. But you will have hopefully realised (and that's the important thing) to what extent both logicist dogmatism and mystical metaphysics extract from this fallacy the illusion of a [nonexistent] source of knowledge—the logicist dogmatists use mere concepts to arrive at their synthetic results, and the intuitive metaphysicians don't even need concepts to get directly to their self-evident axioms.

We are now faced with a paradoxical outcome. We have levelled criticism against both logicist (purely conceptual) metaphysics as well as against mystical (allegedly intuitive) metaphysics, and have recognised the endeavours of both as futile. What is then left? The dilemma that opens in front of us pervades the entire history of philosophy. Indeed, it is the recognition that logicist, scholastic dogmatism is untenable which time and again pushes its critics into the arms of mysticism; and it is the recognition of the impossibility of a philosophy based on a nonexistent intellectual which time and again pulls people back into logicist dogmatism. There doesn't seem to be a way out of this dilemma, unless we count scepticism (the rejection of all metaphysics) as a way out.[12]

The theory I have developed in this course of lectures proves its mettle in that a simple application of it resolves the dilemma. The resolution consists in reducing the dilemma to the same illusion to which we have so far reduced all dialectical illusion—a disjunction whose completeness seems a matter of pure logic. The assumption of completeness is the only thing that gives the dilemma its air of inescapability; you take the assumption away and the dilemma is gone. It is the dilemma between intuition and reflection as possible sources of knowledge.

On the one hand, we have a doubtless logically complete disjunction between reflective and immediate knowledge. On the other hand, we have an equally doubtless logically complete disjunction between intuitive and non-intuitive knowledge. The illusion consists in nothing more than assuming that these two disjunction are one and the same, or to put it more simply, that IMMEDIATE KNOWLEDGE and INTUITION are not two different concepts but just one.

---

[12]This is exactly what happened in Nelson's time with the advent of logical empiricism and logical positivism, an intellectual movement that denied all metaphysical knowledge. Nelson's early death prevented him from confronting that movement. The irony of this story is of course that it was later shown that the famous criterion of demarcation between science and 'metaphysics' cannot possibly belong to science. But all this, as well as the later developments both within analytic and continental philosophy that have led to our contemporary pretty blurred situation, is too large a subject to be tackled here.

The confusion of these two concepts is facilitated if not indeed *forced* by an arbitrary nominal definition. For INTUITION is [nominally and arbitrarily] defined as IMMEDIATE KNOWLEDGE, hence as the opposite of REFLECTIVE KNOWLEDGE or JUDGMENT. However, the opposite to reflective knowledge is rather (from a logical point of view) immediate knowledge, and the question whether immediate knowledge is or isn't intuitive cannot be decided by an arbitrary nominal definition.

We possess a concept, as blurred as you wish, of what we call INTUITION, and it is not a sure thing beforehand that we can define it sufficiently via the concept of IMMEDIATE KNOWLEDGE. A nuanced exploration shows us in fact that INTUITION should not be thus defined, for the ordinary naive way of talking is such that immediate knowledge is called 'intuition' *only if* it is immediately clear to us, so to speak self-evident.[13] Hence saying that *all* immediate knowledge must be intuitive may be right or wrong, but it amounts in any case to passing a synthetic judgment and not an analytic one, as intimated by the above arbitrary definition. Saying that *all* immediate, unreflective knowledge is intuitive, constitutes a synthetic judgment to be proved by something other than just a definition.

Whether the said synthetic judgment is true is a question we don't have to answer here. What is alone important is that the dilemma born from the disjunction of intuition and reflection as the only two possible sources of knowledge is not a matter of pure logic, so that a different kind of inquiry [viz the critique of reason] is needed to see whether in fact a third source of knowledge exists, which consist of neither reflection nor intuition. And so we come to the concept of IMMEDIATE BUT NOT IMMEDIATELY CONSCIOUS KNOWLEDGE. The dilemma and its resolution can be visualised by means of our general schema (Fig. 1).

On the left side of the schema it is stated that we do possess no intellectual intuition, no metaphysical intuition: It is a fact that *our intuition is not intellectual*. To use an example from my last lecture—we have no intuition of the essence of the good but we only bring the good (the law of the good) into our awareness *by thinking*. Because our intuition is not intellectual, we are dragged to logicist dogmatism, for which metaphysics stems from mere reflection. This in turns contradicts the well-known fact placed on the right side of the schema, viz that reflection is empty (for all reflective knowledge is mediated [and comes from somewhere else]). Thus results the correct conclusion, on which the project of mysticism rests, viz that metaphysics cannot originate from reflection, from which the mystics draw the inference that it must originate from a kind of intuition—an intellectual one.

An additional presupposition is common to both parties, i.e. that there is such a thing as metaphysics. But there is another assumption both parties, albeit unconsciously, share, i.e. the very presupposition I have uncovered, viz that the disjunction between intuition and reflection as the only two possible sources of knowledge is complete. It can be stated as follows:

---

[13]Thus at the very least we would have to say that INTUITION is *CLEAR* IMMEDIATE KNOWLEDGE, and not just IMMEDIATE KNOWLEDGE.

Lecture XXII

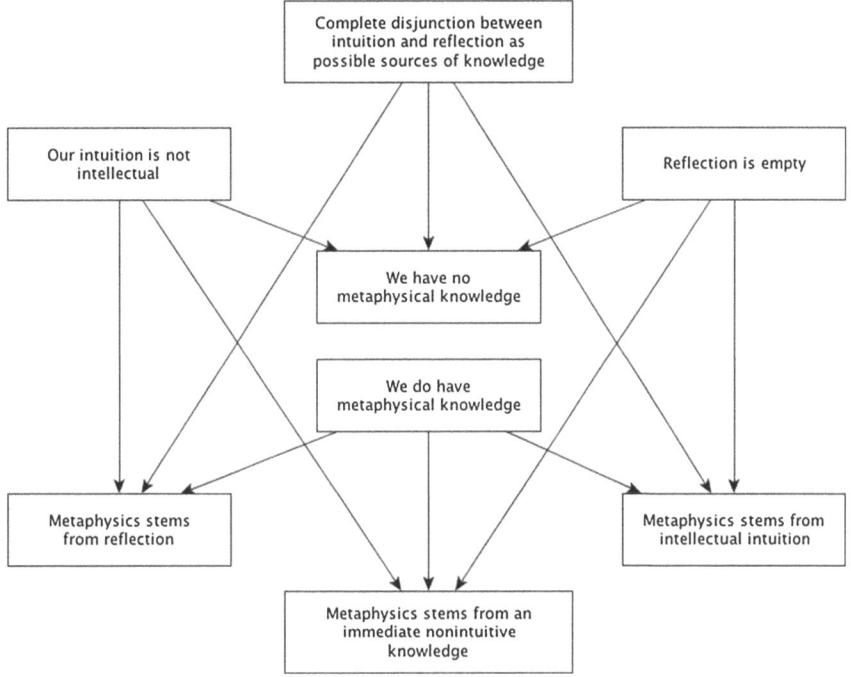

Fig. 1 Empiricists, logicists and intuitionists locked in dispute—and the way out

(P) *All knowledge is either knowledge from intuition or knowledge from experience.*

From (P) and from the fact that we do not possess intellectual intuition it inevitably follows that metaphysics must originate from reflection, if indeed there is such a thing as metaphysics. On the other hand, from (P) and from the fact that reflection is originally empty it follows that metaphysics would have to originate from intellectual intuition if indeed there is such a thing as metaphysics. Finally, from (P) and the two above-mentioned facts follow the desperate position of either scepticism or consistent empiricism, which besides empty reflection only admits sensory intuition as a source of knowledge, and so concludes that we do not possess any metaphysical knowledge at all.

However, once we uncover the dogmatic, not logically validated character of the original disjunction (P), we can without contradiction unite the presuppositions used by the three inferences I just described, thereby maintaining in particular the thesis of the possibility of a metaphysics, albeit one which neither originates from intuition nor from reflection but instead from some non-reflective, i.e. immediate, and yet not intuitive, hence not immediately clear, self-evident knowledge, in short from *some originally obscure immediate knowledge*. To repeat, the illusion rests entirely on saying that the concept of IMMEDIATE KNOWLEDGE is the same as the concept of INTUITION (or IMMEDIATELY CONSCIOUS KNOWLEDGE), and so to consider that the in itself correct disjunction of immediate and reflective knowledge is equivalent to the disjunction of immediately conscious and not immediately conscious knowledge.

That is the resolution of the dilemma. But what happens to metaphysics? The deceptive illusion, the false picture of a transcendental metaphysical knowledge of the world disappears. After all, reflection is empty and our intuition not intellectual. We would have to possess an intellectual intuition in order to advance to a proper metaphysical knowledge of the world. And yet we do possess metaphysical knowledge. We need only establish its restricted scope and content. Critical philosophy, the *critique of reason*, shows that we do possess such knowledge by exhibiting the metaphysical criteria of all knowledge, those criteria that are essentially only guiding lines for experience and all rational induction. Beyond these metaphysical criteria there is nothing left for us but a negative discipline, a critique of the sort elaborated in this course of lectures, a critique of all possible metaphysical errors, albeit a critique which can take on a more satisfying form than the merely negative unmasking of errors, more satisfying because we manage to uncover the most basic explanation of all fallacies and actually arrive at a just appreciation of what metaphysical is made of, thus truly doing it justice. A small contribution to that goal is what I hope to have achieved in these lectures.

References

Bar-Am, Nimrod. 2008. *Extensionalism: The revolution in logic*. Berlin: Springer.
Bencivenga, Ermanno. 2000. *Hegel's dialectical logic*. New York: Oxford University Press.
Frege, Gottlob. 1892. Über Begriff und Gegenstand. *Vierteljahresschrift für wissenschaftliche Philosophie* 16: 192–205. [English translation: On concept and object. *Mind* 60(238): 168–180, 1951].
Fries, Jakob Friedrich. 1828. *Neue, oder anthropologische Kritik der Vernunft*, 3rd ed., 3 vols. Jena. [First edition 1807].
Fries, Jakob Friedrich. 1837. *System der Logik: ein Handbuch für Lehrer und zum Selbstgebrauch* [A system of logic: A handbook for teachers and independent study], 3rd ed. Heidelberg: Winter. [First edition 1811].
Fries, Jakob Friedrich. 1842. Review of *Geschichte der Kant'schen Philosophie* [A history of Kantian Philosophy] by Karl Rosenkranz. *Neue Jenaische allgemeine Literatur-Zeitung* 1 (261–263): 1073–1083.
Hegel, Georg Friedrich Wilhelm. 1821. *Grundlinien der Philosophie des Rechts*. Berlin: Nicolai. [English translation: *Elements of the philosophy of right*. Cambridge: University Press, 1991].
Hilbert, David, and Wolfgang Ackermann. 1928. *Grundzüge der theoretischen Logik*. Berlin: Springer. [English translation of a subsequent edition: *Principles of* mathematical logic. New York: Chelsea, 1950].
Hintikka, Jaakko. 1973. *Logic, language games, and information: Kantian themes in the philosophy of logic*. Oxford: Clarendon Press.
Hintikka, Jaakko. 1996. *La philosophie des mathématiques chez Kant* [Kant's philosophy of Mathematics]. Paris: Presses Universitaires de France.
Husserl, Edmund. 1901. *Logische Untersuchungen*, vol. II. Halle an der Saale: Niemeyer.
Luschei, Eugene C. 1962. *The logical systems of Leśniewski*. Amsterdam: North-Holland.

References

Natorp, Paul. 1910. *Logik (Grundlegung und logischer Aufbau der Mathematik und mathematischen Naturwissenschaften) in Leitsätzen zu akademischen Vorlesungen* [*Logic (Groundwork and logical structure of mathematics and mathematical natural science) in the shape of guidelines for academic courses*]. Marburg: N. G. Elwert.

Nelson, Leonard. 1921. *Spuk — Einweihung in das Geheimnis der Wahrsagerkunst Oswald Spenglers, und sonnenklarer Beweis der Unwiderleglichkeit seiner Weissagungen, nebst Beiträgen zur Physiognomik des Zeitgeistes: eine Pfingstgabe für alle Adepte des metaphysischen Schauens* [*Phantoms—Initiation to the secrets of Oswald Spengler's art of prophecy, plus some contributions to the physiognomy of the spirit of the times: A pentecostal gift for all fans of metaphysical intuition*]. Lepizig: Der neue Geist. [Reprinted in Nelson (1971–1977), vol. III, 349–552].

Nelson, Leonard. 1962. *Fortschritte und Rückschritte der Philosophie von Hume und Kant bis Hegel und Fries*. Edited by Julius Kraft. Hamburg: Felix Meiner [Reprinted in Nelson (1971–1977), vol. VII. English translation: *Progress and regress in philosophy*, 2 vols. Oxford: Blackwell, 1970 and 1971].

Nelson, Leonard. 1971–1977. *Gesammelte Schriften*, 9 vols, eds. Paul Bernays, Willy Eichler, Arnold Gysin, Gustav Heckmann, Grete Henry-Hermann, Fritz von Hippel, Stephan Körner, Werner Kroebel, and Gerhard Weisser. Hamburg: Felix Meiner.

Ringer, Fritz. 1997. *Max Weber's methodology*. Cambridge: Harvard University Press.

Russell, Bertrand. 1903. *The principles of mathematics*, vol. I. Cambridge: University Press [No second volume was ever published].

Spengler, Oswald. 1918. *Der Untergang des Abendlandes: Umrisse einer Morphologie der Geschichte*, vol. 1: *Gestalt und Wirklichkeit*. Munich: Beck. English translation: *The decline of the west: Form and actuality*. New York: Knopf, 1926.

Spengler, Oswald. 1920. *Preussentum und Sozialismus* [*Prussiandom and socialism*]. Munich: Beck.

Urbaniak, Rafal. 2014. *Leśniewski's systems of logic and foundations of mathematics*. Dordrecht: Springer.

Appendix
Seven Kantian Fallacies

[The purpose of this Appendix is to illustrate the fact that Nelson did not think Kant was above committing the kind of fallacies he has described and analysed so thoroughly in his course. The text translated here has been taken from Nelson's *Critical Ethics in the Works of Kant, Schiller and Fries* (1914, 505–519). All footnotes in this Appendix stem from Nelson himself and contain passages selected by him, and mostly quoted *in extenso*, to prove his points. The translations used here come from *Kant's Critique of Practical Reason and Other Works on the Theory of Ethics*, translated by Thomas Kingsmill Abbott (fourth edition revised, London, Longmans, Green & Co., 1889) and available, free of copyright, on the websites of both the Internet Archive and the Gutenberg Project. The reader might want to control the translations by using the Cambridge Edition (see Gregor 1996). The passages quoted by Nelson are marked by bold letters followed by numbers. The letters indicate the particular work the passages are taken from: **G** for *Fundamental Principles of the Metaphysic of Morals*, **K** for the *Critique of Practical Reason*, and **M** for the 'General Introduction' to the *Metaphysics of Morals*. The numbers indicate the pages in the Abbott edition: 1–84 for **G**, 87–262 for **K**, and 265–284 for **M**. In a couple of places I have made slight changes in Abbott's translation. At the end of the description of the fifth fallacy, I omitted a sentence in the original text because it refers the reader to later arguments in Nelson (1914) that are irrelevant to the purpose of this Appendix.—Fernando Leal]

Kant arrived at his discoveries by starting not, like his predecessors, from a dogmatic principle or an arbitrarily stated problem, but from the *fact* of our ethical judgments, and by finding through analysis what the logical presuppositions of those judgments are. The deficiencies of Kant's inquiry begin to appear at the point where he abandons this regressive procedure and thus becomes unfaithful to his own method. This point can be clearly indicated. It has to do with the transition from the *concept* of DUTY to the *law* of duty itself. Kant thought he could derive that law from the sole concept of DUTY instead of finding it by an extended analysis of our ethical judgments.[1] In this transition from the concept of DUTY to the law of duty

---

[1]"In this problem we will first inquire whether the mere conception of a categorical imperative may not perhaps supply us also with the formula of it, containing the proposition which alone can be a categorical imperative. [...] When I conceive a categorical imperative, I know at once what it contains." (**G** 38.) "The categorical imperative, which only expresses in general what obligation is..." (**M** 281.)

a whole series of fallacies occur, which we must consider one by one in order to get rid of the source of the error.

The first fallacy.—In order to bring forth the needed transition, Kant starts from a true proposition: awareness of duty and thus of a practical law is the only reason to act morally, so that judging whether an action is moral must abstract from all content of the law. From that proposition he thinks he can argue that an action's conformity to the law is already a criterion of dutifulness, so that the law of duty should have no content.[2] The mere logical form of the law should suffice to completely determine duty. This is a mistake. First of all, there is no contradiction in assuming that the law has a determinate content whilst saying that nothing beyond the mere form of the law tells me to abide by it. Secondly, this is not only logically possible, it is also necessary. For a practical law whose content is utterly indeterminate is impossible. A command that does not command to do something definite is a contradictory concept. Kant was led to this fallacy by confusing the *criterion* of duty with the *reason* for acting morally.

The second fallacy.—A second connected fallacy allows us to understand how it was that Kant could miss the emptiness of his logicist criterion. 'Duty' is a practical law. So the concept of DUTY entails the concept of UNIVERSAL VALIDITY.[3] What is somebody's duty must be anybody else's duty if placed under the same circumstances. This proposition only makes the universal validity explicit which is contained in the concept of DUTY, yet gives us no criterion as to what one's duty is in a definite situation. For if I only know that I should do what others in the same situation should do as well, then I cannot, without going in circles, decide what a particular person's duty is in a particular situation. The same analytic proposition can also be expressed by saying: 'It cannot be right for somebody to do something that it would not also be right for somebody else in the same situation.' For words like 'right' or 'allowed' is [can be defined as] anything that is not contrary to duty. So the proposition does not say anything more than this: 'If an action carried out by *A* is according to the law, then it must also be according to the law when carried out by *B*.'

---

[2]"But what sort of law can that be, the conception of which must determine the will, even without paying any regard to the effect expected from it, in order that this will may be called good absolutely and without qualification? As I have deprived the will of every impulse which could arise to it from obedience to any law, there remains nothing but the universal conformity of its actions to law in general, which alone is to serve the will as a principle... Here, now, it is the simple conformity to law in general, without assuming any particular law applicable to certain actions, that serves the will as its principle." (**G** 17–18.) "For as the imperative contains besides the law only the necessity that the maxims shall conform to this law, while the law contains no conditions restricting it, there remains nothing but the general statement that the maxim of the action should conform to a universal law." (**G** 38.) See also *Critique of Practical Reason*, Theorems I through III (**K** 107–116).

[3]"From what has been said, it is clear... that it is... of the greatest practical importance,... since moral laws ought to hold good for every rational creature, [to] derive them from the general concept of a rational being." (**G** 28.)

The law itself, which the action must conform to, is still completely undetermined. Without assuming that something else determines what the law is, *any* action would be allowed. For there would be no law to which the action would be contrary.

The illusion that the mere form of the law could serve as criterion of duty is a product of imprecise expression. The popular idea that, say, what is right for anybody is acceptable for anybody else, or the requirement that nobody is privileged over others, does not just mean the truism that what is allowed or forbidden for somebody is allowed or forbidden for everyone else in the same situation; it also means the requirement that when treating people no one should be preferred over others; and in that meaning the proposition *does* serve as a criterion of duty. But it also contains a synthetic judgment that cannot be derived from the mere idea of the universal validity of duty. A 'privilege', such as is excluded by the analytic judgment, would mean allowing someone to do something which is *not allowed* to others placed in the same circumstances. In order to speak of a privilege in this sense we must already have a criterion to decide what is allowed to others. On the contrary, if by 'privilege' we mean allowing someone to do something which we *would not approve* if done by others, then the proposition that no such privileges are possible is a useful criterion of duty. For its application does not presuppose a decision about what other people's duty is or what is right for them to do, but only about what we would want or could not want of others; and that question cannot be decided just by consulting our inclinations, but only by bringing moral principles to bear.

Kant ignored this difference between accepting a course of action as universally commanded (or allowed) and approving that everyone do it. That is why he repeatedly confused the analytic proposition that the maxim of a moral action must have the form of a practical law with the synthetic proposition that it must be possible to *will* it to be a *natural law*.[4] The confusion is facilitated by an imprecise expression Kant was fond of: 'the fitness of a maxim to give a universal law'.[5]

---

[4] "The universal conformity of its actions to law in general,.... i.e. I am never to act otherwise than so that I could also will that my maxim should become a universal law." (**G** 18.) "There remains nothing but the general statement that the maxim of the action should conform to a universal law. [...] There is therefore but one categorical imperative, namely, this: Act only on that maxim whereby thou canst at the same time will that it should become a universal law. [...] Since the universality of the law according to which effects are produced constitutes what is properly called nature in the most general sense (as to form), that is the existence of things so far as it is determined by general laws, the imperative of duty may be expressed thus: Act as if the maxim of thy action were to become by thy will a universal law of nature." (**G** 38–39.) "We must be able to will that a maxim of our action should be a universal law. This is the canon of the moral appreciation of the action generally." (**G** 41.) See also *Critique of Practical Reason*, "Of the typic of the pure practical judgment" (**K** 159–163).

[5] "A practical law which I recognise as such must be qualified for universal legislation; this is an identical proposition." (**K** 115.) It is in fact not even that, but only an analytic proposition that cannot be reversed.

The third fallacy.—A similar fallacy leading to the same result starts from an insufficiently definite concept of AUTONOMY OF THE WILL.[6] Autonomy, the property of not being subject to any other law unless self-imposed by reason, must certainly be considered a property of every rational being. One can say in this sense that the rational will gives itself its own law. It is also true that such a law is binding for *every* rational being, and so that at the same time the will subjects every other rational being to its own legislation. But all those are merely analytic propositions stemming from the concept of PRACTICAL LAW, propositions that are unable to decide anything about the content of such law. Kant was deluded on this point in that he confused autonomy in the sense just defined with the property of a will to be immediately a legislator whose maxims have the status of natural laws, i.e. with the property not to be able to will except so that everybody acts according to the same maxim. Autonomy in this second sense can neither be ascribed to a rational being nor required from it. All that can be required is rather that (another turn of phrase one finds in Kant) a rational being should act *as if* its maxims would amount to universal legislation.[7] The 'principle of a will which in all its maxims gives universal laws'[8] in this sense certainly yields a criterion of duty, but it cannot be derived from the principle of autonomy in the first sense of the word.

The fourth fallacy.—The same fallacy reappears in a slightly different form when Kant tries to derive his proposition of the dignity of the person. That a rational being has dignity (i.e. that it should not be used as a mere means for our ends, but only so that it would itself approve the treatment accorded to it) follows, according to Kant, immediately from the analytic principle of autonomy that a rational being

---

[6]"Autonomy of the will is that property of it by which it is a law to itself (independently of any property of the objects of volition). The principle of autonomy then is: *Always so to choose that the same volition shall comprehend the maxims of our choice as a universal law.*" (**G** 59.) "Autonomy, that is to say, the capability of the maxims of every good will to make themselves a universal law, is itself the only law which the will of every rational being imposes on itself." (**G** 63.) "If there is a categorical imperative (i.e. a law for the will of every rational being), it can only command that everything be done from maxims of one's will regarded as a will which could at the same time will that it should itself give universal laws." (**G** 50.)
[7]"The formula of the moral imperative is expressed thus, that the maxims must be so chosen as if they were to serve as universal laws of nature." (**G** 54.) "Therefore every rational being must so act as if he were by his maxims in every case a legislating member in the universal kingdom of ends. The formal principle of these maxims is: *So act as if thy maxim were to serve likewise as the universal law (of all rational beings).* [...] But although a rational being, even if he punctually follows this maxim himself, cannot reckon upon all others being therefore true to the same,... still that law: *Act according to the maxims of a member of a merely possible kingdom of ends legislating in it universally,* remains in its full force, inasmuch as it commands categorically." (**G** 57.)
[8]**G** 50.

cannot be subjected to any law that it does not give itself.[9] It is true that, according to this principle of autonomy, a rational being certainly cannot depend on any alien will *to tell it what its duty is*. Yet in relation to the *content* of such duty nothing has thereby been decided. Hence it does not contradict that [analytic] autonomy when a rational being is subjected to the ruthless whim of another and thus used by it as a mere means. Kant confuses here the autonomy of rational beings as subject of duties with their autonomy as objects of duties, i.e. he confuses the property of a rational being, *as a universal legislator*, to restrict the will of other rational beings with its claim to have its *ends* or *interests* respected by them.

Speaking more precisely only the law itself can be said to restrict the will. Yet insofar as every rational being imposes a law on itself by the use of its reason, one can also say that it restricts the will of all other rational beings, for these also, being rational, are subjected to the same law. On the other hand, insofar as the content of the law is to restrict the will of every rational being so as to respect the interest of all beings it deals with, one can also express this briefly by saying that the will of every rational being is restricted by the interest of all beings it deals with. Autonomy, as the property of being self-legislating and thereby restricting the will of all rational beings, only belongs to a being insofar as it has *reason* and is itself the *subject of duties*. Yet autonomy as dignity, i.e. as the property of restricting, through its interests, the will of all rational beings, does not belong to a being insofar as it has reason but insofar as it has interests and so is *subject of rights*. The synthetic proposition of the dignity of the person, i.e. the property a being has of restricting, through its interests, the will of all rational beings, cannot therefore be derived from the analytic proposition of the autonomy of rational beings…

---

[9]"Reason then refers every maxim of the will, regarding it as legislating universally, to every other will… from the idea of the dignity of a rational being, obeying no law but that which he himself also gives." (**G** 53.) "Autonomy then is the basis of the dignity of human and of every rational nature." (**G** 54.) "The principle: *So act in regard to every rational being (thyself and others), that he may always have place in thy maxim as an end in himself*, is accordingly essentially identical with this other: *Act upon a maxim which, at the same time, involves its own universal validity for every rational being*. For that in using means for every end I should limit my maxim by the condition of its holding good as a law for every subject, this comes to the same thing as that the fundamental principle of all maxims of action must be that the subject of all ends, i.e., the rational being himself, be never employed merely as means, but as the supreme condition restricting the use of all means, that is in every case as an end likewise." (**G** 56.) "He must always take his maxims from the point of view which regards himself and, likewise, every other rational being as law-giving beings (on which account they are called persons)." (**G** 57.) "In all creation every thing one chooses and over which one has any power, may be used merely as means; man alone, and with him every rational creature, is an end in himself. By virtue of the autonomy of his freedom he is the subject of the moral law, which is holy. Just for this reason every will, even every person's own individual will, in relation to itself, is restricted to the condition of agreement with the autonomy of the rational being, that is to say, that it is not to be subject to any purpose which cannot accord with a law which might arise from the will of the passive subject himself; the latter is, therefore, never to be employed merely as means, but as itself also, concurrently, an end." (**K** 180–181.)

The fifth fallacy.—Side by side with his derivation of the proposition of the dignity of the person, Kant tries to argue for it in different ways. One way is his conception of the absolute value of the good will. He starts by saying that the good will is the only good not subject to any restricting condition, and from that he argues that the rational being, as subject of a possibly good will, is alone valuable in an absolute sense and so it is an object of respect.[10] We are again entangled in a logical circle here. First, if this argument is right, then in treating a person we would have the duty to take into account only its moral ends and not its morally indifferent ones. Secondly, affirming that the person has absolute value directly contradicts the principle that the good will is the only unconditional good. But apart from these two points, having defined the good will as the will to respect the law, we cannot without a logical circle now define the law as commanding us to respect the good will.

The fallacy is already at work in trying to derive the absolute value of the good will. The will is good because it subjects itself to the command of duty. But duty is practical necessity, and to fulfill it is what we ought to do, and does not carry any merit. Therefore, no positive value can be produced by doing one's duty.[11] Sure, we give preference to the action that is our duty over any other action; and so it has a *comparatively* infinite value. This is part of the concept of DUTY. But from that we cannot argue that fulfilling one's duty has any absolute value; all we can conclude is that not fulfilling it has an absolute disvalue. The good will is therefore certainly the necessary condition of all personal value, but this does not mean that it is itself a positive good, never mind the highest.

The sixth fallacy.—In the arguments considered so far, Kant has obviously been concerned to avoid the fallacy of an ethics of goods. Instead of basing the moral law upon an absolute value, he has always remained faithful to the proposition that

---

[10] "Whatever has a price can be replaced by something else which is equivalent; whatever, on the other hand, is above all price, and therefore admits of no equivalent, has a dignity. [...] Now morality is the condition under which alone a rational being can be an end in himself, since by this alone is it possible that he should be a legislating member in the kingdom of ends. Thus morality, and humanity as capable of it, is that which alone has dignity." (**G** 53.) "An independently existing end... can be nothing but the subject of all possible ends, since this is also the subject of a possible absolutely good will." (**G** 56.) "This ideal will which is possible to us is the proper object of respect." (G 59.) "The worth of a character perfectly accordant with the moral law is infinite." (**K** 225.)

[11] See **K** 175, as well as **K** 153 ("the supreme condition of all good"), **K** 166 ("the first condition of personal worth"), **K** 171 ("that freedom... restricts the estimation of the person by the condition of obedience to its pure law"), **K** 181–182 ("it is the effect of a respect for something quite different from life, something in comparison and contrast with which life with all its enjoyment has no value"), **G** 20 ("an estimation of the worth which far outweighs all worth of what is recommended by inclination"), **G** 54 ("this estimation therefore shows that the worth of such a disposition is dignity, and places it infinitely above all value, with which it cannot for a moment be brought into comparison or competition without as it were violating its sanctity"), **G** 69 ("worth, in comparison with which that of an agreeable or disagreeable condition is to be regarded as nothing", "a worth simply in our own person which can compensate us for the loss of everything that gives worth to our condition").

Appendix: Seven Kantian Fallacies

'nothing has any value but the one assigned to it by the law'.[12] Kant has so far only overreached himself in starting from the *concept* of LAW instead of starting from the law itself. Yet another line of argument he pursues reverses the relation between law and value. He tries to establish the absolute value of the person and then to base the law upon that value.[13] This argument avoids the logicist fallacy but only at the expense of falling back into an ethics of goods.

The proof of the absolute value of the person which Kant had in mind rests on the proposition that the value of things is conditioned—by the inclinations of a person that are to be satisfied using those things as means. Since to any conditioned value there must be an unconditional one which is its condition of possibility, yet the inclinations themselves (as sources of needs) cannot obviously have such an unconditional value, then it indeed seems as though we had an argument leading to the absolute value of the person.[14]

This illusion, however, stems from the ambiguity of the word 'conditioned'. That a value is conditioned means, on the one hand, that it depends from some other value, to which it is related as a means to an end. In this sense of the word we can of course say that the value of things is only conditioned, and we can also say that every conditioned value presupposes an unconditioned one that guarantees its possibility. But if we look for the unconditioned value in this sense, the value that makes any value of things possible, then we do not arrive at the value of the *person*, but at the value of the satisfaction of needs or the wellbeing—a *state*—of a person.

That a value is conditioned means, on the other hand, that it depends on a *valuing subject*—the property an object can only have to the extent that it satisfies the inclination of a needy being.[15] In this sense of the word we can rightly say that the person is the condition of any value that any object of inclinations can have, but we could never say that all conditioned value presupposes an unconditional value to

---

[12] **G** 54. See also **K** 154 ("that the concept of good and evil must not be determined before the moral law—of which it seems as if it must be the foundation—but only—as it is done here—after it and by means of it"), **K** 155 ("that it is not the concept of good as an object that determines the moral law and makes it possible, but that, on the contrary, it is the moral law that first determines the concept of good and makes it possible, so far as it deserves the name of good absolutely").

[13] "Supposing, however, that there were something whose existence has in itself an absolute worth, something which, being an end in itself, could be a source of definite laws; then in this and this alone would lie the source of a possible categorical imperative, i.e., a practical law." (**G** 46.) "If all worth were conditioned and therefore contingent, then there would be no supreme practical principle of reason whatever. If then there is a supreme practical principle or, in respect of the human will, a categorical imperative, it must be one which, being drawn from the conception of that which is necessarily an end for everyone because it is an end in itself, constitutes an objective principle of will, and can therefore serve as a universal practical law. The foundation of this principle is: rational nature exists as an end in itself." (**G** 46–47.)

[14] **G** 46.

[15] "All objects of the inclinations have only a conditional worth, for if the inclinations and the wants founded on them did not exist, then their object would be without value." (**G** 46.)

guarantee its possibility. There is no argument in this sense that would lead us to the absolute value of the person. Kant has confused the relation between mediated and immediate values with the relation between subjective and objective values.[16]

The seventh fallacy.—But never mind this fallacy. We are still missing the transition from the absolute value of the person to the command 'to respect the person as an end in itself'. The argument is like this. Call *P* the proposition that the person has absolute value or is an end in itself. *P* has two possible meanings. On the one hand, *P* can mean that the person should not be made into a mere means.[17] But then *P* is just another way of expressing the categorical imperative, not a higher principle to which we can reduce it. On the other hand, *P* could express a law of value *V* that does not presuppose the categorical imperative. But then the categorical imperative cannot be derived from *V*, for we cannot argue from the absolute value of the person to saying that the person is an object of duty unless we presuppose that it is our duty to bring about what is absolutely valuable or (if this is not in our will's power) to preserve it. But in that case the highest command of duty would be this presupposition, and not the one we were trying to derive, viz 'to respect the dignity of the person'. So there is no way we can derive the law of duty from the absolute value of the person.

Kant's illusion that such a derivation is possible despite everything is strengthened by another confusion. Kant equates the contrast between conditioned and unconditioned values (in the different senses we just discussed) with the contrast between replaceable (finite) and irreplaceable (infinite) values.[18] However, even if we take the latter, the absolute value of the person does not suffice to make it the object of duty. For we would still have to go from the infinite value of the object of an action to that of the action itself. And even the infinite value of the action would not suffice to recognise it as our duty. This is so because, even if we would leave aside the negative character of all moral value, even if we would say that a dutiful action has infinite value, the reverse cannot be argued—that an infinitely valuable action is our duty.

---

[16]"Beings whose existence depends not on our will but on nature's, have nevertheless, if they are irrational beings, only a relative value as means, and are therefore called *things*; rational beings, on the contrary, are called *persons*, because their very nature points them out as ends in themselves... These, therefore, are not merely subjective ends whose existence has a worth *for us* as an effect of our action, but *objective ends*, that is, things whose existence is an end in itself." (**G** 46.)

[17]"As ends in themselves, that is as something which must not be used merely as means." (**G** 46.) "As ends, that is, as beings who must be capable of containing in themselves the end of the very same action." (**G** 48.)

[18]"Whatever has a price can be replaced by something else which is equivalent; whatever, on the other hand, is above all price, and therefore admits of no equivalent, has a dignity... But that which constitutes the condition under which alone anything can be an end in itself, this has not merely a relative worth, i.e., value, but an intrinsic worth, that is, *dignity*." (**G** 53.)

References

Gregor, Mary J., ed. and transl. 1996. *Immanuel Kant: Practical philosophy.* New York: Cambridge University Press.

Nelson, Leonard. 1914. Die kritische Ethik bei Kant, Schiller und Fries: eine Revision ihrer Prinzipien [Critical ethics in the works of Kant, schiller, and fries: A revision of principles]. *Abhandlungen der Fries'schen Schule (N.F.)* 4(3): 483–691 [Reprinted in Nelson (1971–1977), vol. VIII, pp. 27–192].

Nelson, Leonard. 1971–1977. *Gesammelte Schriften,* 9 vols. Edited by Paul Bernays, Willy Eichler, Arnold Gysin, Gustav Heckmann, Grete Henry-Hermann, Fritz von Hippel, Stephan Körner, Werner Kroebel, and Gerhard Weisser. Hamburg: Felix Meiner.

The manufacturer's authorised representative in the EU is Springer Nature Customer Service Centre GmbH, Europaplatz 3, 69115 Heidelberg, Germany. If you have any concerns regarding our products, please contact ProductSafety@springernature.com

Printed and bound by CPI Group (UK) Ltd, Croydon, CR0 4YY

23/03/2026

02076668-0012